Principles of Electronic Communication Systems

Principles of Electronic Communication Systems

Contributors

Pekka JÄNIS, Chia-Hao YU et al.

AURIS
Reference

www.aurisreference.com

Principles of Electronic Communication Systems

Contributors: Pekka JÄNIS, Chia-Hao YU et al.

Published by Auris Reference Limited

www.aurisreference.com

United Kingdom

Principles of Electronic Communication Systems

ISBN:978-1-78154-924-7

British Library Cataloguing in Publication Data
A CIP record for this book is available from the British Library

Printed in the United Kingdom

Exclusively distributed by CBS Publishers & Distributors Pvt. Ltd.

Sales & Distribution Rights only for India, Pakistan, Bangladesh, Sri Lanka, Nepal and Bhutan. This book is not to be sold outside these territories.

Contents

List of Abbreviations

ADC	Analog-to-Digital Converter
AOA	Angle of Arrival
ANN	Artificial Neural Network
AFC	Automatic Frequency Control
AGC	Automatic Gain Control
BLE	Bluetooth Low Energy
CAD	channel allocation and deallocation
CC	channel changing
CDN	channel deciding node
CDMA	Code Division Multiple Access
CAP	Contention Access Period
CFP	Contention Free Period
D2D	Device-to-Device
EFA	Exploratory Factor Analysis
EKF	Extended Kalman Filter
FIR	Finite impulse response
GPS	Global Positioning System
GTS	Guaranteed Time Slots
HCP	host control platform
IC	integrated circuit
IPM	integrated pest management
ITS	Intelligent Transportation Systems
IMM	Interacting Multiple Model
IF	Intermediate Frequency
LMS	Least-mean-squares
LOS	Line-of-sight
LTE	Long Term Evolution
MAE	Mean absolute error
MAC	medium access control
MSA	Micro Signal Architecture
MCA	Multi-Channel Access
MDS	Multidimensional Scaling
MPC	Multi-path components
NSF	National Science Foundation
NLD	network load detection
OSM	Open Street Map
PDP	Parallel Distributed Processing
PLE	Path loss exponent
PIC	Peripheral Interface Controller
PDP	Power delay profile
PFC	power factor corrector

QoS	Quality of Service
RLG	Research Libraries Group
SNR	Signal-to-Noise Ratios
TOA	Time of Arrival
UWB	Ultra wideband
UTC	Universal Time Coordinated
V2V	Vehicle-to-vehicle
VIL	virtual induction loop
VCO	Voltage Controlled Oscillator
WD	wave digital
WSN	Wireless Sensor Network

List of Contributors

Pekka JÄNIS
Department of Signal Processing and Acoustics, Helsinki University of Technology, Espoo, Finland

Chia-Hao YU
Department of Communications and Networking, Helsinki University of Technology, Espoo, Finland

Klaus DOPPLER
Nokia Research Center, Nokia Group, Helsinki, Finland

Cássio RIBEIRO
Nokia Research Center, Nokia Group, Helsinki, Finland

Carl WIJTING
Nokia Research Center, Nokia Group, Helsinki, Finland

Klaus HUGL
Nokia Research Center, Nokia Group, Helsinki, Finland

Olav TIRKKONEN
Department of Communications and Networking, Helsinki University of Technology, Espoo, Finland

Visa KOIVUNEN
Department of Signal Processing and Acoustics, Helsinki University of Technology, Espoo, Finland

Rondia MACK
Department of Electrical and Computer Engineering, Indiana University Purdue University Indianapolis, Indianapolis, USA

Maher RIZKALLA
Department of Electrical and Computer Engineering, Indiana University Purdue University Indianapolis, Indianapolis, USA

Paul SALAMA
Department of Electrical and Computer Engineering, Indiana University Purdue University Indianapolis, Indianapolis, USA

Mohamed EL-SHARKAWY
Department of Electrical and Computer Engineering, Indiana University Purdue University Indianapolis, Indianapolis, USA

ANNE J. GILLILAND-SWETLAND
Council on Library and Information Resources 1755 Massachusetts Avenue, NW, Suite 500 Washington, DC 20036

GREG KINNEY
Council on Library and Information Resources 1755 Massachusetts Avenue, NW, Suite 500 Washington, DC 20036

Ruizhen Han
College of Biosystems Engineering and Food Science, Zhejiang University, Hangzhou 310058, China
College of Electronic Information, Zhejiang University of Media and Communications, Hangzhou 310018, China

Yong He
College of Biosystems Engineering and Food Science, Zhejiang University, Hangzhou 310058, China

Fei Liu
College of Biosystems Engineering and Food Science, Zhejiang University, Hangzhou 310058, China

Md. Shariful Islam
Department of Computer Engineering, Kyung Hee University, 1 Seocheon, Giheung, Yongin, Gyeonggi 449-701, Korea

Muhammad Mahbub Alam
Department of Computer Engineering, Kyung Hee University, 1 Seocheon, Giheung, Yongin, Gyeonggi 449-701, Korea

Choong Seon Hong
Department of Computer Engineering, Kyung Hee University, 1 Seocheon, Giheung, Yongin, Gyeonggi 449-701, Korea

Sungwon Lee
Department of Computer Engineering, Kyung Hee University, 1 Seocheon, Giheung, Yongin, Gyeonggi 449-701, Korea

Marco Gramaglia
Institute IMDEA Networks, Avenida del Mar Mediterraneo 22, 28918 Leganes (Madrid), Spain
Department of Telematics Engineering, Universidad Carlos III de Madrid, Avda. Universidad, 30, 28911 Leganes (Madrid), Spain

Carlos J. Bernardos
Department of Telematics Engineering, Universidad Carlos III de Madrid, Avda. Universidad, 30, 28911 Leganes (Madrid), Spain

Maria Calderon
Department of Telematics Engineering, Universidad Carlos III de Madrid, Avda. Universidad, 30, 28911 Leganes (Madrid), Spain

Youhei Kawamura
Faculty of Engineering, Science and Information Systems, University of Tsukuba, Ibaraki Prefecture 305-8573, Japan

Markus Wagner
School of Computer Science, University of Adelaide, Adelaide, SA 5005, Australia

Hyongdoo Jang
Department of Mining Engineering and Metallurgical Engineering, Curtin University, Perth, WA 6433, Australia

Hajime Nobuhara
Faculty of Engineering, Science and Information Systems, University of Tsukuba, Ibaraki Prefecture 305-8573, Japan

Takeshi Shibuya
Faculty of Engineering, Science and Information Systems, University of Tsukuba, Ibaraki Prefecture 305-8573, Japan

Itaru Kitahara
Faculty of Engineering, Science and Information Systems, University of Tsukuba, Ibaraki Prefecture 305-8573, Japan

Ashraf M Dewan
Department of Spatial Sciences, Curtin University, Perth, WA 6433, Australia

Bert Veenendaal
Department of Spatial Sciences, Curtin University, Perth, WA 6433, Australia

Yousif I. Al Mashhadany
Electrical Engineering Department, Engineering College, University of Anbar, Baghdad, Iraq

Kazi Nazmul Huda
Department of Business Administration, Southern University Bangladesh, Chittagong, Bangladesh

Abul Kalam Azad
Department of Communication & Journalism, University of Chittagong, Chittagong, Bangladesh

Javier Portela
Group of Analysis, Security and Systems (GASS), Department of Software Engineering and Artificial Intelligence (DISIA), Faculty of Information Technology and Computer Science, Office 431, Universidad Complutense de Madrid (UCM), Calle Profesor José García Santesmases, 9, Ciudad Universitaria, Madrid 28040, Spain

Luis Javier García Villalba
Group of Analysis, Security and Systems (GASS), Department of Software Engineering and Artificial Intelligence (DISIA), Faculty of Information Technology and Computer Science, Office 431, Universidad Complutense de Madrid (UCM), Calle Profesor José García Santesmases, 9, Ciudad Universitaria, Madrid 28040, Spain

Alejandra Guadalupe Silva Trujillo
Group of Analysis, Security and Systems (GASS), Department of Software Engineering and Artificial Intelligence (DISIA), Faculty of Information Technology and Computer Science, Office 431, Universidad Complutense de Madrid (UCM), Calle Profesor José García Santesmases, 9, Ciudad Universitaria, Madrid 28040, Spain
Facultad de Ingeniería, Universidad Autónoma de San Luis Potosí (UASLP), Zona Universitaria Poniente, San Luis Potosí 78290, Mexico

Ana Lucila Sandoval Orozco
Group of Analysis, Security and Systems (GASS), Department of Software Engineering and Artificial Intelligence (DISIA), Faculty of Information Technology and Computer Science, Office 431, Universidad Complutense

de Madrid (UCM), Calle Profesor José García Santesmases, 9, Ciudad Universitaria, Madrid 28040, Spain

Tai-hoon Kim
Department of Convergence Security, Sungshin Women's University, 249-1 Dongseon-dong 3-ga, Seoul 136-742, Korea

P. Visconti
Department of Innovation Engineering, University of Salento, 73100 Lecce, Italy

S. D'Amico
Department of Innovation Engineering, University of Salento, 73100 Lecce, Italy

A. Baschirotto
Department of Physics, University Bicocca, 20126 Milano, Italy

D. Romanello
Electronic Division, Cavalera Sistemi s.r.l., Galatone, 73044 Lecce, Italy

P. Costantini
Department of Innovation Engineering, University of Salento, 73100 Lecce, Italy

V. Ventura
Department of Innovation Engineering, University of Salento, 73100 Lecce, Italy

G. Cavalera
Electronic Division, Cavalera Sistemi s.r.l., Galatone, 73044 Lecce, Italy

Juan Chóliz
Research Institute of Engineering in Aragón, I3A, University of Zaragoza, C/ María de Luna 3, Zaragoza 50018, Spain

Ángela Hernández
Research Institute of Engineering in Aragón, I3A, University of Zaragoza, C/ María de Luna 3, Zaragoza 50018, Spain

Antonio Valdovinos
Research Institute of Engineering in Aragón, I3A, University of Zaragoza, C/ María de Luna 3, Zaragoza 50018, Spain

Robert Nagel
Institute of Communication Networks, Technische Universität München Germany

Stefan Morscher
Institute of Communication Networks, Technische Universität München Germany

Preface

Principles of Electronic Communication Systems offers the most up-to-date coverage of the rapidly changing communications field. An electronic communication is generally an electronic system that widely disseminates orders entered by market makers to third parties and permits the orders to be executed against in whole or in part. In first chapter, we propose to facilitate local peer-to-peer communication by a Device-to-Device (D2D) radio that operates as an underlay network to an IMT-Advanced cellular network. Second chapter addresses the feasibility of lowering the clock frequency of the processing unit that models the PLL is addressed and modulator/demodulator functions of the system while maintaining synchronization with the memory unit and other peripherals. Third chapter provides an overview of computer conferencing systems; in particular the CONFER I1 software used at the University as well as many other academic and non-academic institutions, and then outlines the methodology and major findings of the project. The chapter also discusses some issues raised by the project that may well be generic to the archival management of electronic communication. Fourth chapter presents a feasibility study on a real-time in field pest classification system design based on Blackfin DSP and 3G wireless communication technology. This prototype system is composed of remote on-line classification platform (ROCP), which uses a digital signal processor (DSP) as a core CPU, and a host control platform (HCP). In fifth chapter, we present a multi-channel communications system for WSNs that is referred to as load-adaptive practical multi-channel communications (LPMC). In sixth chapter, we propose and evaluate VIL (Virtual Induction Loop), a simple and lightweight traffic monitoring system based on cooperative vehicular communications. Seventh chapter focuses on a multimedia data visualization based on ad hoc communication networks and its application to disaster management. Eighth chapter describes the implementation of a PIC microcontroller in a conventional laboratory-type electronic trainer. The key objective of ninth chapter is to identify the major determinants of professional stress of journalists in electronic media. Tenth chapter aims to develop a global statistical disclosure attack to detect relationships between users. Eleventh chapter presents the bidirectional power line communication system developed in parallel to an electronic board for driving and control of HID (high-intensity discharge) and LED (light-emitting diode) lamps. Twelfth chapter provides a complete system level evaluation of a UWB-based communication and location system for Wireless Sensor Networks, including aspects such as UWB-based ranging, tracking algorithms, latency, target mobility and MAC layer design. Last chapter highlights on the connectivity prediction in mobile vehicular environments backed by digital maps.

Chapter 1

DEVICE-TO-DEVICE COMMUNICATION UNDERLAYING CELLULAR COMMUNICATIONS SYSTEMS

Pekka Jänis[1], Chia-Hao Yu[2], Klaus Doppler[3], Cássio Ribeiro[3],
Carl Wijting[3], Klaus Hugl[3], Olav Tirkkonen[2], Visa Koivunen[1]

[1]Department of Signal Processing and Acoustics, Helsinki University of Technology, Espoo, Finland

[2]Department of Communications and Networking, Helsinki University of Technology, Espoo, Finland

[3]Nokia Research Center, Nokia Group, Helsinki, Finland

ABSTRACT

In this article we propose to facilitate local peer-to-peer communication by a Device-to-Device (D2D) radio that operates as an underlay network to an IMT-Advanced cellular network. It is expected that local services may utilize mobile peer-to-peer communication instead of central server based communication for rich multimedia services. The main challenge of the underlay radio in a multi-cell environment is to limit the interference to the cellular network while achieving a reasonable link budget for the D2D radio. We propose a novel power control mechanism for D2D connections that share cellular uplink resources. The mechanism limits the maximum D2D transmit power utilizing cellular power control information of the devices in D2D communication. Thereby it enables underlaying D2D communication even in interference-limited networks with full load and without degrading the performance of the cellular network. Secondly, we study a single cell scenario consisting of a device communicating with the base station and two devices that communicate with each other. The results demonstrate that the D2D radio, sharing the same resources as the cellular network, can provide higher capacity (sum rate) compared to pure cellular communication where all the data is transmitted through the base station.

INTRODUCTION

Major effort has been spent in recent years on the development of next-generation wireless communication systems that will bring higher data rates and system capacity to end users and network operators. Examples of such next-generation systems are 3GPP Long Term Evolution (LTE) and WiMAX (see http://www.3gpp.org/ and http: //www.wimaxforum.org/).

Currently the evolution of such systems has been started under the scope of IMT-Advanced. In addition to traditional performance targets of high data rates and better coverage, the success of IMT-Advanced systems will depend on their ability to enable new services. It is expected that local services will contribute significantly to the growth of mobile communications. The widespread development of local services will be enabled by decreasing infrastructure costs and direct connectivity that supports peer-to-peer communication between local services and the end users. In fact already today mobile phones act as web server (see http://mymobilesite.net/) and offer direct connectivity, e.g. using Bluetooth technology.

In this article we propose to facilitate the local peer-topeer communication by a Device-to-Device (D2D) radio that operates as an underlay network to an IMT-Advanced cellular network. This D2D radio is a potential key enabler for low cost, seamless and high capacity local connectivity. We assume the infrastructure network to be a cellular network based on Orthogonal Frequency Division Multiple Access (OFDMA) technology. The cellular network operates in licensed bands, and it is important to guarantee that D2D transmissions will not generate harmful interference to cellular users. Similar problems are observed in the context of cognitive radios [1-3], where the cellular usage is the primary service.

In order to control the interference from D2D connections to the cellular network, we propose that the Base Station (BS) is able to control the maximum transmit power and the resources to be used for each D2D connection. Note that such a scenario is different from pure ad-hoc networks, without coordination from an infrastructure network, e.g. [4,5]. Further, we present a novel power control mechanism for D2D connections that share cellular uplink resources. The mechanism limits the maximum D2D transmit power, utilizing cellular power control information of the devices in D2D communication. The performance of D2D and cellular communications is evaluated by means of system simulations that include interference from multiple cells.

Secondly, we study a single cell scenario consisting of a device communicating with the BS and two devices that communicate with each other. We consider three modes of operation: D2D communication can share either uplink (UL) or downlink (DL) resources with the cellular network or use exclusive resources. If direct communication between the terminals is not beneficial, the two devices communicate through the BS of the cellular network. In semi-analytical studies we show that the D2D radio, sharing the same resources as the cellular network, can provide higher capacity (sum rate) than pure cellular communication through the BS.

This article is organized as follows: In Section 2 we present the motivation for mobile D2D communications with an example application and give a brief overview of the state of the art in D2D communication. In Section 3 we present the power control mechanism for the coexistence of D2D and cellular transmissions. In Section 4 we describe the simulation methodology and present the simulation results. In Section 5 the semi-analytical analysis is described and results are presented. In Section 6 we present results on indoor D2D connections sharing the DL resources with a metropolitan area network. In Section 7 we summarize our results and the conclusions are given.

MOBILE DEVICE-TO-DEVICE COMMUNICATION

Next generation mobile communication systems such as 3GPP LTE and WiMAX are optimized for wide area and metropolitan area operation. In recent years local area networks based on WLAN have been increasingly popular, as they enable access to the internet and to local services with low cost APs and cheap and fast access to wireless spectrum in the license exempt bands. However only a licensed band can guarantee a controlled interference environment and local service providers might prefer to pay a small amount of money to get access to licensed spectrum when the license exempt bands get crowded. Cellular operators may offer such cheap access to spectrum with controlled interference enabled by D2D communication as underlay to the cellular network.

This concept is illustrated in **Figure 1**, where UE denotes User Equipment. The BS allows UE2 and UE3 to communicate directly to each other while keeping some control over the D2D link to limit the interference to the cellular receiver. As an example, consider the case where a media server is put up at a rock concert from which visitors can download promotional material using the D2D connection. At the same time, the cellular network can handle phone calls and internet data traffic without the additional load that would be caused by traffic from the media server. The D2D operation itself can be transparent to the user. She simply enters a URL, the network would detect traffic to the media server and hand it over to a D2D connection. The same application could also be enabled by a media server with built in WLAN AP or Bluetooth. However in that case the user has to define the WLAN AP or perform Bluetooth pairing which can be tedious especially if a secure connection is required.

Compared to other local connectivity solutions based on for example Bluetooth or WLAN the D2D communication supported by a cellular network offers additional compelling advantages. First the network can advertise local services available within the current cell. Thus for automated service discovery, the devices do not have to constantly scan for available WLAN AP or Bluetooth devices. This is especially advantageous when considering that the constant scanning of Bluetooth devices or WLAN APs is often switched off to reduce the power consumption. Secondly, the cellular network can distribute encryption keys to both D2D devices so that a secure connection can be established without manual pairing of devices or entering encryption keys.

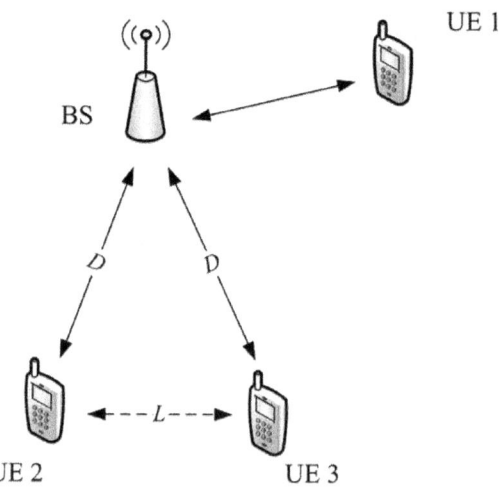

Figure 1: Illustration of D2D communication as an underlay network to an infrastructure network. UE1 is a cellular user whereas UE2 and UE3 are D2D users. D denotes the distance between D2D nodes and BS, and L denotes the D2D link distance.

Several wireless standards have addressed the need for D2D operation in the same band as the BS, also called Access Point (AP) or central controller. Examples of such standards are Hiperlan 2 [6], TETRA [7], and WLAN.

In all these standards D2D communication is assumed to occur on separate resources. For example, in standards employing Orthogonal Frequency Division Multiplexing (OFDM) as the physical layer, like Hiperlan 2, User Equipments (UE) involved in D2D communications are not allowed to share the same OFDM symbol with UEs communicating to the infrastructure network. This restriction limits the interference. However it leads to inefficient utilization of resources, especially for large system bandwidths. The resource utilization is even more inefficient in TETRA, where several frequency channels are country-wide reserved solely for D2D communication; reducing the resources available for cellular communication when no D2D communication is present.

In WLAN the UE senses the medium and transmits if the resources are free. A drawback of such a scheme is that the AP does not have a direct possibility of controlling the D2D links and providing assistance, which could prove highly beneficial for the network [8].

D2D POWER CONTROL WHEN SHARING UL RE-SOURCES

Since the D2D communication takes place as an underlay communication to the cellular OFDMA network, the interference from D2D communication to the cellular network has to be coordinated and the BS should be aware of ongoing D2D connections. The UEs in D2D connections are still associated to the BS and can receive for example cellular calls. Thus, we propose that the D2D link initialization and the allocation of OFDMA Resource Block (RB) to the D2D links is managed by the BS. Therefore, there is an immediate opportunity for the BS to reduce the interference between the cellular and D2D links. Such a scheme for sharing UL resources is proposed in this section. The power level of D2D transmitters is chosen based on the cellular UL power control information to limit the interference to the cellular BS.

The easiest way to restrict the D2D interference would be to mandate a predefined maximum power level to the D2D transmitters, and this level could be chosen such that the expected degradation in the cellular links stays at a tolerable level. However, such an approach would have to be designed for the worst case scenario and would lead to inefficient use of resources. As the D2D transmitter may be arbitrarily close to the cellular receiver, the power level thus determined is likely to be inadequate for establishment of reliable D2D links, other than for extremely short range communication. On the other hand, the power level of the D2D transmitter could be substantially higher if the network would have some means to determine how close the D2D transmitter is to the cellular receiver.

In fact the cellular BS has just the required information for controlling the interference from D2D transmitters to the cellular BS in case the D2D links share UL resources with the cellular network. The UL power control in a cellular network aims at reducing the dynamic range of signals received from multiple devices, i.e. to reduce the near-far effect. The BS may use the cellular UL power control framework in setting the D2D transmit power. To be more specific, let us consider the SINR of the UL cellular transmission in an isolated cell with ideal transceivers and flat fading channel. In this case the expression for the cellular UL SINR may be written as

$$\xi = \frac{P_1 c_1}{P_2 c_2 + \sigma_w^2}$$

(1)

where P_1 and P_2 denote the transmit powers of the cellular and D2D UEs, c_1 and c_2 the corresponding link gains to the base station, and σ_w^2 the additive white Gaussian noise power. The base station has full control over the powers P_1 and P_2 in Equation (1), given the limitations on the transmit power range of the terminal and on the dynamic range of the power control specified for the radio interface.

Equation (1) implies that in case of ideal UL power control without the presence of a D2D transmitter (P_2=0) and a target SNR of P/σ_w^2, the cellular power control target is $P_1 c_1 = P$.

In order to keep the interference to UL transmissions under control, the BS can signal the D2D transmitter to apply a power level such that $P_2 c_2 = P/B$, where B is a backoff parameter. For large values of the backoff parameter B, D2D transmissions cause very low interference to UL transmissions. However, a large B implies a reduced range for the D2D link itself. We can avoid this limitation on the range of the D2D link by incorporating a power boosting factor α to the transmit power of the UL transmitter that compensates for the remaining interference from D2D transmissions. In this case, Equation (1) is modified to

$$\xi' = \frac{\alpha P_1 c_1}{P_2 c_2 + \sigma_w^2}$$

(2)

The power boosting value α is defined such that the received SINR from UL transmission in Equation (2) is equal to the target SNR, i.e. $\xi' = P/\sigma_w^2$. Hence, substituting $P_1 c_1 = P$ and $P_2 c_2 = P/B$ into Equation (2), we obtain

$$\alpha = \frac{P}{B\sigma_w^2} + 1$$

(3)

Naturally, in case of no UL transmissions, the D2D transmitter does not need to apply a power backoff, i.e. $B=1$. Conversely, no power boosting is needed in case of no D2D transmissions. In fact, in cases when only a subset of the RBs is used by D2D traffic, the power boosting is only applied to those UL transmissions that share the RBs with a D2D pair. As a result, the received UL power is non-uniform over the system bandwidth, which tends to increase the inter-RB interference caused by power amplifier nonlinearities and limited receiver dynamic range. Moreover, UL transmissions are not perfectly orthogonal due to the effects of non-ideal synchronization and wireless propagation environment. Therefore we limit the boosting values to 10dB, which we assume to be still manageable.

NUMERICAL RESULTS ON COEXISTENCE OF D2D COMMUNICATION AND CELLULAR NETWORK

In this section we study the coexistence of D2D communication links with an interference-limited cellular local area network. The D2D pair is sharing either UL or DL resources with the cellular links. The aim is to find out the achievable D2D link quality when giving priority to the cellular links. The study is carried out by static system simulations and empirical SINR distributions for both the cellular and D2D links are evaluated.

Scenario and Channel Model

The scenario and network layout resembles a local area indoor scenario, illustrated in **Figure 2**. Nine BS serve a whole floor of 100m times 100m. The scenario incorporates small room, corridor like longer rooms and a large open area in the center. Similar elements can be typically found in shopping malls or office areas.

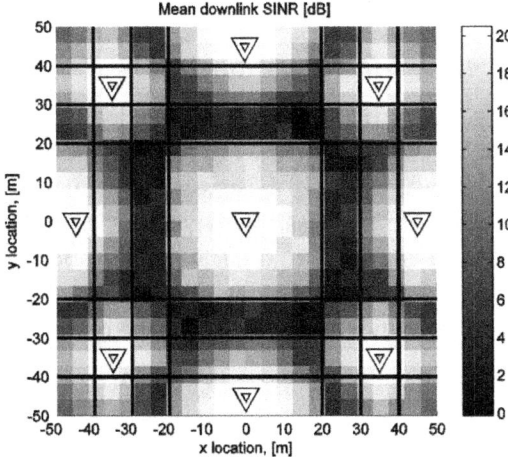

Figure 2: The simulated scenario for coexistence studies. The triangles represent the locations of base stations and the black horizontal and vertical lines represent walls.

The color indicates the mean DL SINR without D2D interference as a function of location.

We have used the channel and propagation models defined in WINNER [9] scenario A1 (indoor/office) for our studies. Links in the same room have a distance dependent probability for Line-Of-Sight (LOS) conditions.

In the channel model, for the LOS and Non-LineOf-Sight (NLOS) propagation conditions, the path-loss exponent is 1.87 and 3.68, respectively, and the shadow fading standard deviation is 3dB and 4dB, respectively. In addition, each wall introduces an additional attenuation of 5dB. Frequency selective fading is also modeled, with a resolution of 2MHz.

The cellular UEs are uniformly distributed over the area and the locations of the D2D pairs are independent from the cellular UEs. The D2D pairs are generated with the restriction that the D2D link must reside within a single room.

System Model

We assume that the network operates on a 100MHz band using Time Division Duplexing (TDD). The base stations have acquired frame-synchronism and use the same split between UL and DL resources, such that there is no interference from neighboring cell DL transmission to UL transmissions or vice versa. The modulation scheme allows Frequency Division Multiplexing (FDM) transmissions from the BS to several UEs simultaneously, as well as Frequency Division Multiple Access (FDMA) for several UEs to the BS. Specifically, the 100MHz band is split into five orthogonal RBs of 20MHz.

Scheduler for Cellular Transmissions

Each BS randomly selects five UEs to be scheduled, from those UEs that are not in D2D mode. For each UE it tries to allocate the RB with best SNR or the RB with second best SNR. If these are not available, the first free RB is allocated. In case there are less than five UEs associated to the BS, the remaining free RBs are allocated to a UE with allocation in the adjacent RB. Hence, the network is fully loaded at every time instant. Since the channel is assumed reciprocal, the same frequency resources that are used for UL are also used for DL.

UL Power Control

For each BS, the total transmitted power is 25dBm, which is evenly distributed over all sub-bands, i.e. 18dBm for each 20MHz band. For the UE, the uplink power control aims a target SNR of P/σ_w^2, limited by the maximum transmit power of 18dBm for the UE. With these settings, in the studied scenario about 10% of the UEs utilize maximum output power. When the UL power boosting defined in Section 3 is used, the portion of UEs reaching maximum transmit power increases to 40%.

SINR Calculations

The SINR of each 2MHz sub-band for a transmission originating from node n and received by node m is calculated as

$$\xi_{n,m} = \frac{P_n c_{n,m}}{\sigma_w^2 + \sigma_{Tx}^2 + \sigma_{Rx}^2 + I_{NL} + \sum_{k \neq n} P_k c_{k,n}}$$

where I_{NL} is the power of all out-of-band emissions at the receiver, and σ_w^2, σ_{Tx}^2, and σ_{Rx}^2 are the thermal noise power, the transmitter in-band distortion, and the receiver Analog-to-Digital Converter (ADC) noise floor, respectively, integrated over the 2MHz sub-bands. The thermal noise power is derived from the noise figure defined in LTE specifications for UE and BS [10]. With the chosen transmit power levels our network scenario is clearly interference-limited.

Network Simulation Description

The simulation arrangement is as follows. A large number of random independent snapshots of network operation, called drops, are modeled. For each drop, a set of cellular and D2D UEs are independently placed in the scenario. The path-loss and shadow fading values of each link are then determined and each cellular and D2D UE is associated with the BS to which it has the strongest link. Each cell has 5 cellular UEs and 5 D2D pairs, and the D2D transmissions are multiplexed to 5 sub-bands. This way we can ensure that each cellular transmission is interfered by exactly one intracell D2D link and vice versa. After all transmissions are scheduled, the transmit powers of the D2D transmitters are determined as in Section 3 and the SINR is computed as in Subsection 4.2.3.

Results and Discussion

In this section we present the simulation results on the SINR distributions for the cellular UL, cellular DL, and the D2D links. A wide range of transmit power levels without power control for the D2D link was simulated along with an UL power control based D2D case. The settings for the power control case have been chosen such that for 95% of the cellular links the SINR degradation is less than 3dB. This was achieved with a power backoff of $B=5$ dB, for a UL power control target SNR of 13dB. From Equation (3), this implies an UL power boost $\alpha=8.64$ dB.

From **Figure 3**(a) and 3(b), we observe that the maximum allowed D2D transmit power should be limited to -10dBm in DL and -24dBm in the UL phase of the frame, assuming that a 3dB degradation of the cellular SINR at the 5-th percentile of the SINR CDF would be still tolerated. Assuming as well that a D2D link with $SINR \geq 0$ dB is usable, we observe from 3(c) and 3(d) that the fraction of usable D2D links is ≈45% in DL phase and ≈33% in UL phase.

However, if the UL-based D2D power control scheme is applied, the percentage of usable links rises to 73% in UL phase while still maintaining the same cellular performance as for -24dBm D2D transmit power, as observed in **Figure 3**(b) and 3(d). As it can be appreciated from **Figure 3** and from the discussion above, when the D2D transmit power is set to a fixed level such that the degradation to the cellular performance remains tolerable, D2D performance is slightly better when it uses DL resources than when it uses UL resources. This may sound counter-intuitive since, due to the fact that the BS's constant transmit power is significantly higher than the mean UL transmit power, the interference to the D2D links is higher in the DL than in UL phase. On the other hand, due to the same reason the cellular DL transmissions can tolerate much higher D2D transmit powers than the cellular UL transmissions. In the UL phase, the D2D transmit power must be set such that even in the event of a D2D transmitter being close to the BS the cellular performance does not degrade too much. Since UL power control guarantees that all UEs experience similar SINR in their UL transmissions regardless of their position in the network, this becomes a very strict requirement.

The proposed UL-based power control scheme effectively removes this restriction, resulting in significant improvement on the D2D performance. A similar power control scheme is not applicable in DL since the BS might schedule resources shared with D2D links to multiple cellular UEs. Each of these candidate cellular UEs would have to set a power control target to the D2D transmitters, implying significant overhead.

RESOURCE ALLOCATION ANALYSIS

In the preceding sections we considered the coexistence of a D2D underlay and cellular network. Specifically, we demonstrated that it is possible to allow D2D communication to share the cellular resources and at the same time guarantee that the performance of the cellular communication links is not sacrificed, thus taking an approach where the cellular communication has priority. In this section we take a different viewpoint where neither the cellular nor the D2D communication have priority over the other. This gives insight on the maximum benefits in terms of overall performance that D2D underlay communication can provide.

We assume the Channel State Information (CSI) of all the involved links is available at BS so that the resource allocation decision of D2D users can be controlled centrally by the BS. The scenario where at most one cellular user and one D2D communication pair will share the same radio resource is considered. Despite its simplicity, this scenario captures the minimum requirement for cellular communication and D2D communication to share the same resource.

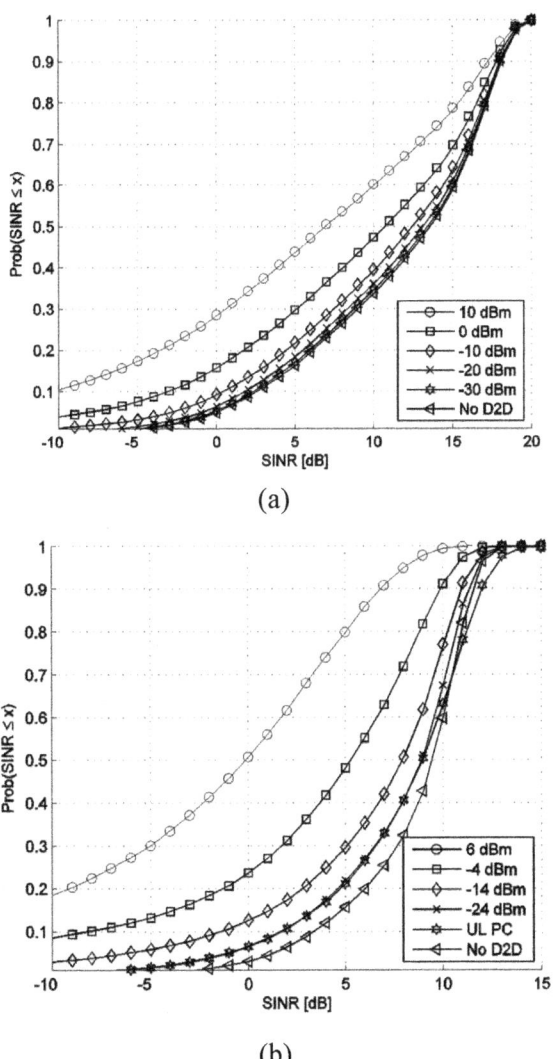

(a)

(b)

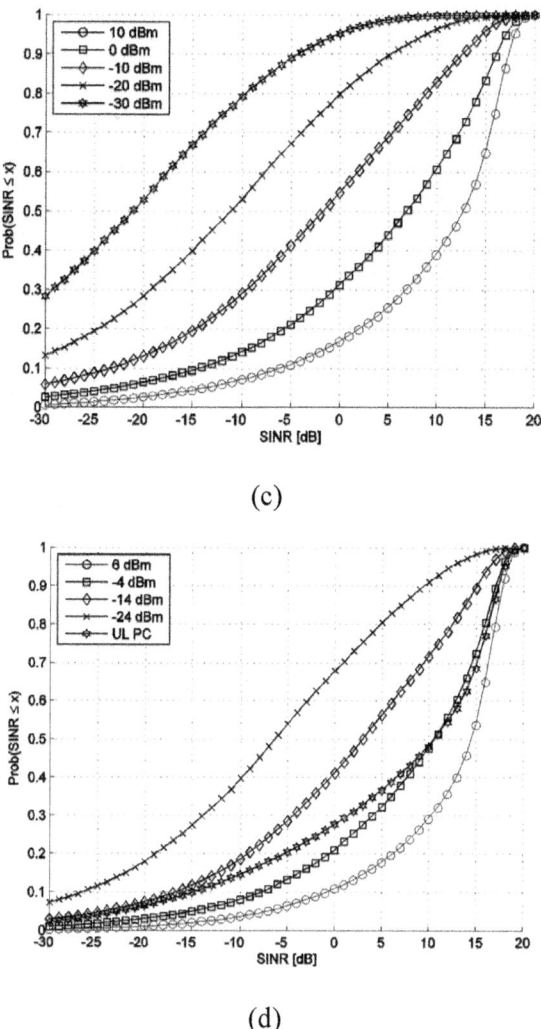

(c)

(d)

(a) DL SINR. (b) UL SINR.(c) D2D SINR in DL phase.(d) D2D SINR in UL phase.

Figure 3: Empirical SINR CDFs in the local area scenario with various D2D transmitter power levels. The UL and DL SINR without D2D is shown for reference. The D2D SINR distributions are shown conditioned on the link distance being less than 8 meters. Each D2D link is within a single room.

In general, D2D communication causes no interference to the cellular users if they occupy separate resources. However, resource usage efficiency can be higher if the same resource is shared at the same time. We may achieve higher overall system performance if D2D and cellular communications co-exist in the same radio resource. We will discuss four different resource allocation modes including both separate and non-separate sharing schemes. They are illustrated in **Figure 4** and detailed below:

- DL resource sharing (DLre): D2D communication happens in DL resources so that all the DL resources of the cellular user are interfered.

- UL resource sharing (ULre): Similar to DLre, D2D communication happens in UL resources, and all the UL resources of the cellular user are interfered.

- Separate resource sharing (SEPre): D2D communication takes half of the available resources from the cellular user, either from DL or UL resource. There is no interference between cellular and D2D communication.

- Cellular mode sharing (CellMod): The D2D users communicate with each other through the BS that acts like a relay node. They take half of the available resources either from the DL or the UL resources of the cellular user. Note that this mode is conceptually the same as traditional cellular system and is used as a reference.

In the following, we consider a normalized isolated circular cell (with radius equal to 1) as illustrated in

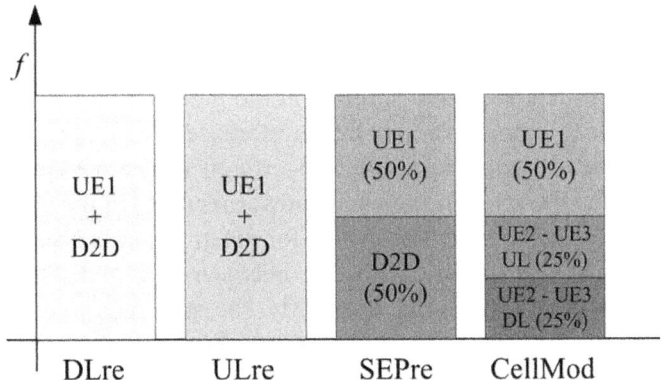

Figure 4: Illustration of resource allocation of considered resource sharing methods. In DLre and ULre modes, the cellular user and the D2D users operate in the same resource. In SEPre and CellMod modes, the cellular user and the D2D users occupy different resources.

Figure 1, and discuss the best resource allocation scheme out of the four possible modes. We assume one cellular user (UE1) and two D2D users (UE2

and UE3) sharing the available radio resources. For simplicity, we consider only distance-dependent pathloss, but no fading. Specifically, we consider the single-slope pathloss model [11] with pathloss exponent 4:

$$P(d) = \frac{P(d_0)}{d^4}$$

(4)

where $P(d)$ denotes the received power at the distance d from the transmitter and $P(d_0)$ is the received power at reference distance d_0. To adapt the normalized cell considered in our environment, we simply replace $P(d_0)$ with the transmit power. This channel model enables a one-to-one mapping between the distance of a channel link and the received signal strength. In addition, since the considered channel model provides the mean channel condition in a fading channel, the trend presented in this simplified model is consistent with the case where a more complex model is applied. We assume the distance between the D2D users and the BS to be D and the distance between the two D2D users to be L. Assuming no power control, the transmit power is fixed to unity for the cellular transmission in UL and DL, and for the D2D user transmissions. Note that the per-RB transmit powers used in Section 4 are according to this scheme-the BS and the UE have the same (maximum) power spectral density. Under this channel model and the geometric constraint of the D2D users, the resource allocation decision depends on the cellular user (UE1) position only, under a given set of D and L.

The interference caused by D2D users may come from any D2D users depending on which one is transmitting at the moment. Here, we assume the worst interference condition where the interference from D2D communication is caused by the user creating stronger interference. The AWGN noise power is assumed to be the same as the signal power received at the cell border (i.e. SNR=0dB at the cell edge). The metric for determining the resource sharing mode is the sum rate of the connection between UE2 and UE3, and of the cellular connection between BS and UE1. The sum rate takes into account either DL or UL resources depending on which one is shared with the D2D users. The sum rate is calculated by the Shannon capacity formula [12] according to the following equations

$$R_{\text{ULre}} = \log_2\left(1 + \frac{P_{23}}{\max(P_{12}, P_{13}) + N_0}\right)$$
$$+ \log_2\left(1 + \frac{P_1}{\max(P_2, P_3) + N_0}\right)$$

$$R_{\text{DLre}} = \log_2\left(1 + \frac{P_{23}}{\max(P_2, P_3) + N_0}\right)$$
$$+ \log_2\left(1 + \frac{P_1}{\max(P_{12}, P_{13}) + N_0}\right)$$

$$R_{\text{SEPre}} = \frac{1}{2}\log_2\left(1 + \frac{P_{23}}{N_0}\right) + \frac{1}{2}\log_2\left(1 + \frac{P_1}{N_0}\right)$$

$$R_{\text{CellMod}} = \frac{1}{2}\cdot\frac{1}{4}\left(\log_2\left(1 + \frac{P_2}{N_0}\right) + \log_2\left(1 + \frac{P_3}{N_0}\right)\right)$$
$$+ \frac{1}{2}\log_2\left(1 + \frac{P_1}{N_0}\right),$$

where P_i denotes the received power of the link between BS and UE i, and P_{ij} denotes the received power of the link between UE i and UE j. N_0 is the noise power at the receiver. The received power in each link is calculated by Equation (4).

The resource allocation scheme which gives the best sum rate is selected for each UE1 position according to

$$R_{\max} = \max\left(R_{\text{ULre}}, R_{\text{DLre}}, R_{\text{SEPre}}, R_{\text{CellMod}}\right) \tag{5}$$

It should be noted that, without any further constraint, it may happen that either the cellular or the D2D connection is compromised in order to maximize the sum rate. For example, under the condition that UE1 is very close to BS, it is likely that the connection between the BS and the UE1 dominates the selection of the resource allocation scheme. The selected scheme might give the D2D connection little transmission rate. Similarly, when the D2D users are very close to each other and dominate the sum rate, the transmission rate of UE1 may be very limited. **Figure 5** shows the share of cell area where one specific resource allocation scheme is selected as the best one, under different values of D and L. The curves corresponding to CellMod mode are missing because it is selected only under the condition that the two D2D users are at opposite sides with respect to BS $(L \approx 2D)$. In this special condition, the CellMod mode is the favorable resource sharing scheme. This can be observed from **Figure 5** by noticing that the share of all curves with $L = 0.6$ goes to approximately 0 at $D = 0.3$. However, except for this particular case, D2D communication is always beneficial for the system.

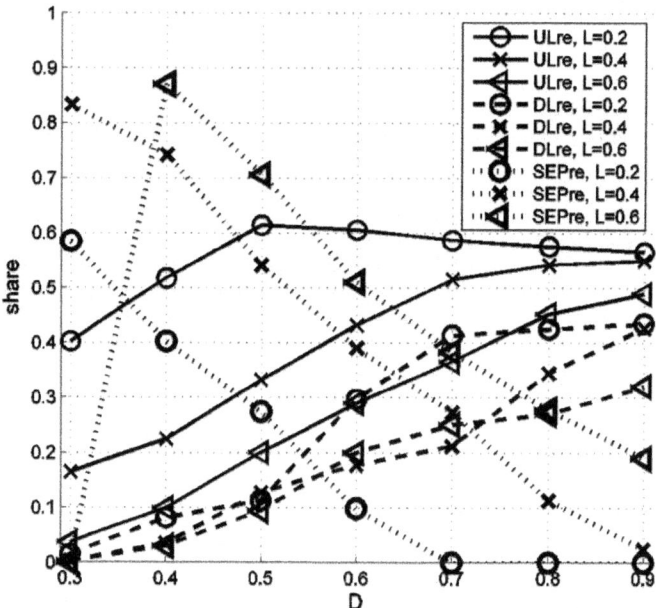

Figure 5: Share of cell area where one specific resource allocation scheme is selected as the best one, for different values of D (the distance between BS and the D2D pair) and L (the distance between the two D2D users). D2D communication is always beneficial for the system except for the case when the D2D users are at opposite sides with respect to BS (i.e. D=0.3 and L=0.6), where none of ULre, DLre and SEPre resource allocation schemes occupies significant percentage area.

When the D2D users are further away from BS (i.e. when D is large), the percentage area where the cellular user experiences strong interference reduces. It suggests the benefit of using non-separate resource sharing schemes (ULre or DLre) which provide higher resource usage efficiency. When D is small, it is more beneficial to use either ULre or SEPre depending on the value of L. In small D and small L scenario, the signal strength between the D2D users is very strong. The ULre mode outperforms DLre mode in this case because the interference observed by the D2D users is smaller in ULre mode, which significantly improves channel quality of D2D communication.

Figure 6 shows the rate ratio of D2D communication to the CellMod mode under the parameter $D = 0.7$ and $L = 0.2$. The two circular spots give the position of two D2D users. The cellular user is at one given position in the cell, and the color at that position represents the rate gain from D2D communication, which is the rate ratio between the rates obtained from the best resource sharing scheme and the CellMod mode.

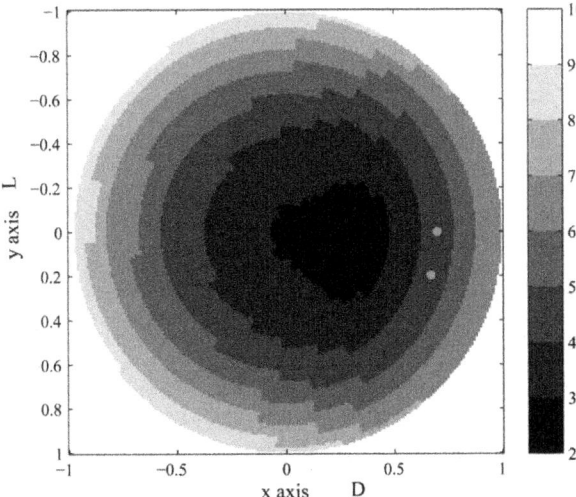

Figure 6: Rate ratio to CellMod mode with D=0.7 and L=0.2. The two circular spots denote the position of D2D users. The cellular user is at a given position in the cell and the background color at the position displays the rate ratio of the best resource sharing scheme to the CellMod mode.

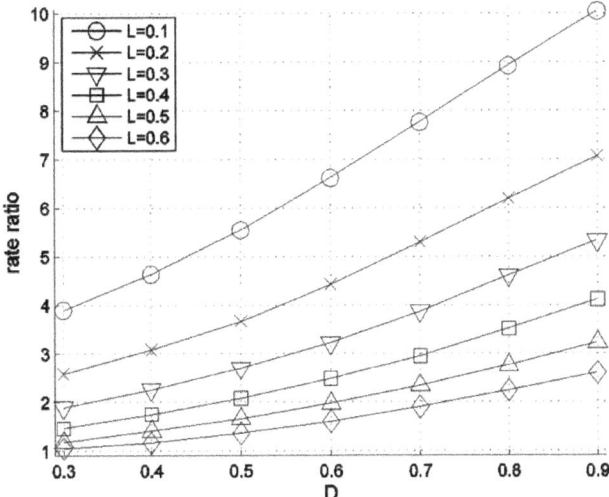

Figure 7: Average rate ratio over CellMod. The rate ratio is obtained by comparing the rate obtained from the best resource sharing scheme and the rate obtained from the CellMod mode. For different values of D and L, the rate gain is averaged over the whole cell.

The area outside the unit circle is out of the considered cell and should not be considered. The gain is significant and depends on the position of the cellular user. In **Figure 7**, we illustrate the rate gain averaged over the considered single-cell with respect to different geometry of D2D users. It is clear that, in average, the larger D and the smaller L are beneficial for the system performance. Consistent with **Figure 5**, the rate gain vanishes when the D2D users are at opposite sides of BS (e.g. $D = 0.3$ and $L = 0.6$).

INDOOR D2D AS UNDERLAY TO A METROPOLITAN-AREA NETWORK

In the previous sections we have demonstrated that D2D communication can take place in an interference-limited network as well as the potential gains from D2D communication in a single cell. Now we consider D2D as an underlay to a metropolitan area network.

The cellular BS are deployed outdoors and outdoor cellular users share downlink resources with indoor D2D connections. The BSs are deployed in a multi-cell environment and the results are obtained from the center cell. The BS deployment is modelled by the well-known Manhattan grid and follows the UMTS 30.03 recommendation [13] and the corresponding channel and path-loss models can be found in [9].

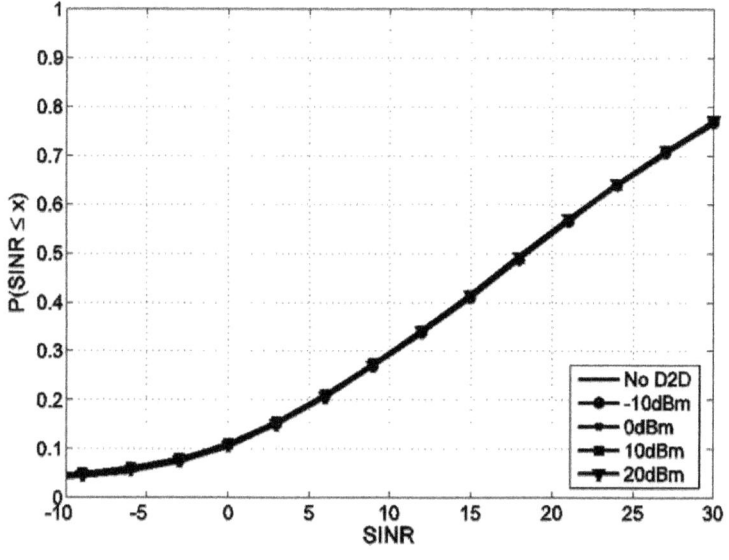

(a)

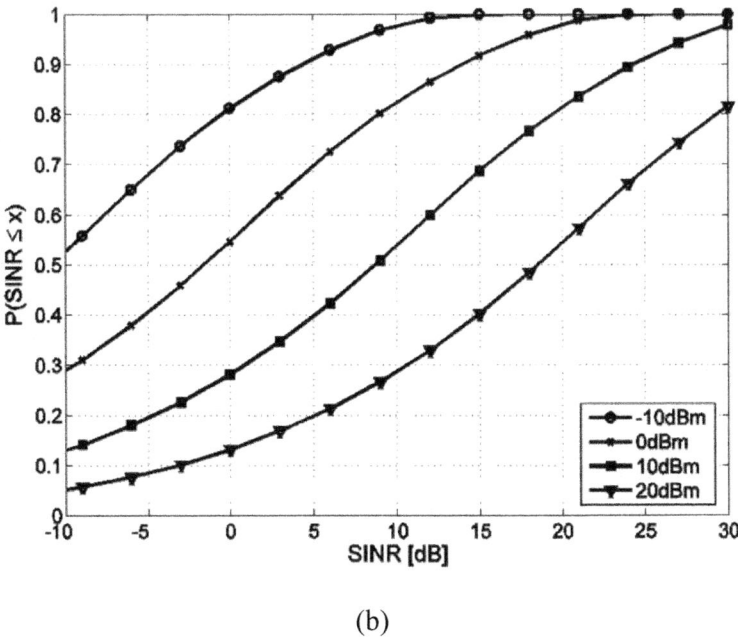

(b)

(a) Empirical CDF of cellular SINR for different D2D transmission power. There is no visible impact from the D2D communication on the SINR of the cellular network. (b) Empirical CDF of D2D connection SINR for different D2D transmission power.

Figure 8: SINR of cellular and D2D connections sharing the cellular downlink resources.

The D2D pairs are randomly generated within the same building block and the pathloss between them is below 90dB which corresponds to a distance of up to 25m. The penetration loss of at least 14dB through the outside wall of the building and the favorable propagation between outdoor BS and cellular devices in the same street isolates the indoor D2D connections from the outer cellular network. Both the cellular devices and the D2D devices operate with full buffers, i.e. both the cellular network and the D2D devices utilize the full bandwidth with 100% load. A single OFDMA resource block (RB) is shared by one cellular user and one D2D pair in the cell.

The transmit power of the cellular BS is set to 37dBm and the transmit power of the D2D devices was limited to 20dBm. **Figure 8** illustrates the potential for D2D connection with different transmit power in such a scenario. The cellular SINR is not affected by the indoor D2D connections even when they transmit with 20dBm. About 90% of the D2D connections experience a higher SINR than 0dB which we see as a lower threshold where D2D communication makes sense. In general the BS will also serve indoor users and the example might be overly optimistic. Nevertheless the BS can for example allocate only part of the downlink resources to D2D connections and schedule only outdoor cellular users in these resources. The BS can for example classify outdoor users based on path-loss, spatial signature or location information.

CONCLUSIONS

In this paper we analyze Device-to-Device (D2D) communications underlaying a cellular network. We show that given proper power control and coordination mechanisms it is possible to have D2D connections that reuse cellular band and still cause only minimal interference to the cellular network.

We propose a power control scheme for the D2D links that share uplink resources with a cellular network. In this case the maximum power that can be used for the D2D link is defined by taking the cellular uplink power control information as reference. We evaluate the proposed power control scheme in system simulations. The results show that by properly defining the maximum power on the D2D link a good D2D link SINR is achieved while at the same time the impact on the cellular network is minor. Thereby D2D communication can take place in interference-limited networks with full load, where a cognitive radio would not be able to detect a white space.

Further, we performed semi-analytical studies on a single-cell scenario to analyze how much gain can be expected from D2D communications. We considered several allocation strategies, including traditional cellular communications. The results show that significant gains in sum rate can be achieved by enabling D2D communications compared to the conventional cellular system.

Finally, we showed in system simulations that indoor D2D communication causes negligible interference to outdoor cellular users in the downlink of a metropolitan area network.

REFERENCES

1.   S. Haykin, "Cognitive radio: Brain-empowered wireless communications," IEEE Journal on Selected Areas in Communications, Vol. 23, No. 2, pp. 201-220, February 2005.

2.   J. Mitola and G. Q. Maguire Jr., "Cognitive radio: Making software radios more personal," IEEE Personal Communications, Vol. 6, No. 4, pp. 13-18, August 1999.

3.   I. F. Akyldiz, W. -Y. Lee, M. C. Vuran, and S. Mohanty, "Next generation

dynamic spectrum access cognitive radio wireless networks: A survey," Computer Networks: The International Journal of Computer and Telecommunications Networking, Vol. 50, No. 13, pp. 2127-2159, September 2006.

4.   C. -K. Toh, M. Delwar, and D. Allen, "Evaluating the communication performance of an ad hoc wireless network," IEEE Transactions on Wireless Computing, Vol. 1, No. 3, pp. 402-414, July 2002.

5.   Y. Xue, B. Li, and K. Nahrstedt, "Optimal resource allocation in wireless ad hoc networks: A price-based approach," IEEE Transactions on Mobile Computing, Vol. 5, No. 4, pp. 347-364, April 2006.

6.   ETSI, "BRAN; HIPERLAN2 type 2; data link control (DLC) layer; part 4: Extension for home environment," TS 101 761-4, v1.3.2, 2002.

7.   ETSI, "Terrestrial trunked radio (TETRA); voice plus data (V+D) designers' guide; part 3: Direct mode operation (DMO)," TR 102 300-3 v1.2.1, 2002.

8.   H.-Y. Hsieh and R. Sivakumar, "On using peer-to-peer communication in cellular wireless data networks," IEEE Transactions on Mobile Computing, Vol. 3. No. 1, pp. 57-72, January-February 2004.

9.   WINNER II D1.1.2, "WINNER II channel models," https:// www.ist-winner.org/deliverables.html, September 2007.

10.   [3GPP TR 25.814 V7.1.0, "Technical specification group radio access network; physical layer aspects for evolved universal terrestrial radio access (UTRA)," http://www. 3gpp.org/, September 2006.

11.   T. S. Rappaport, "Wireless communication principles and practice," New Jersey: Prentice Hall, 1996.

12.   T. M. Cover and J. A. Thomas, "Elements of information theory," New York: Wiley, 1991.

13.   ETSI, "Recommendation TR 30.03 selection procedure for the choice of radio transmission technologies of the UMTS," 1997.

Chapter 2

VLSI IMPLEMENTATION FOR LOW NOISE POWER EFFICIENCY CELLULAR COMMUNICATION SYSTEMS

Rondia Mack, Maher Rizkalla, Paul Salama, Mohamed El-Sharkawy

Department of Electrical and Computer Engineering, Indiana University Purdue University Indianapolis, Indianapolis, USA

ABSTRACT

A low power model for Code Division Multiple Access (CDMA) based cellular communication system is developed. The dynamic power is minimized by reducing the frequency of the Phase Lock Loop (PLL) after lock is established. The paper addresses the feasibility of lowering the clock frequency of the processing unit that models the PLL is addressed and modulator/demodulator functions of the system while maintaining synchronization with the memory unit and other peripherals. The system is simulated with Matlab considering various Signal-to-Noise Ratios (SNR). For a given SNR, the minimum frequency required for the PLL to maintain lock is determined. The Matlab file is translated to VHDL code, simulated and synthesized with Mentor tools, and the layout then generated. Mach-Pa 5-V software system from Mentor tools is utilized to estimate the power consumed by the simulated device. A Xilinx file is also generated and downloaded for Field Programmable Gate Arrays (FPGA) implementation. A 50 MHz clock frequency of the processing unit was first considered and then lowered to 20 MHz for the low power study. Lowering the base and clock frequency resulted in near 30% reduction in power.

INTRODUCTION

Cellular phone systems require a large number of base stations in each city regardless of the size. An average large city can have hund reds of towers. In addition, each carrier in each city also runs one central office called the

Mobile Telephone Switching Office (MTSO). This office handles all of the cellular phone connections to the land- based phone system and controls all of the base stations in the region. An on-chip PLL (Phase Lock Loop) gener-ates the internal clock at one of 16 frequencies ranging from 88 to 287 MHz based on a fixed 3.68 MHz input clock [1]. It is a system requirement that the chip return quickly from the idle state to normal operation with no such constraint on returning from the sleep state. Based on this determination and the 20 mW power budget in Idle, it was concluded that when th e PLL power is below 2 mW then the PLL can run in Idle and remove the re-quirements on the PLL lock time. Thus, there is need for a very low power PLL dictated by the power budget in Idle [2]. The TDA8012M is a low power PLL FM demodulator for satellite TV receivers [3].

It supports low power be-cause it has a sensitive PLL FM demodulator, and is used for the second Intermediate Frequency (IF) filter in satel-lite receivers. It also provides Automatic Gain Control (AGC) and Automatic Frequency Control (AFC) outputs that can be used to optimize the level and frequency of the input signal. During the searching procedure, the AFC output provides a signal which is used for carrier detection, high input sensitivity, and balanced two-pin Voltage Controlled Oscillator (VCO) and Carrier detec-tor. Low power filtering was researched b y the digit-serial implementation method of all pass filter structure [4]. Also the general-order lossless and discrete integrator/ differentiator methods were involved. In low-power-filter implementation, digit-serial computation showed to be advantageous compared to bit-serial and parallel arithmetic [5]. The digital-serial processing elements are obtained using unfolding techniques. The implementa-tion is compared to a corresponding wave digital (WD) all pass filter implementation [6]. For low power con-sumption, ASIC and FPGA design architectures were implemented for modulation/demodulation, Chebyshev scheme was used for filtering, QPSK for transmission, and CDMA/TDMA for channelization.

APPROACH

Power reduction techniques include pipelining, parallelism, reducing clock cycles, and lowering the frequency that will also raise the stability issue for several electronic components. In this paper, the PLL is one of the main sources to reduce power in the cell phone system. The values of I (t) and Q (t) represent the binary digits in QPSK format. The S_{input} values represent the modulated output signal from the base station. Two mixers are needed for the Q (t) and I (t) variables, where $Q(t) = \cos(2\pi ft + j)S_{input}$, $I(t) =) = \cos(2\pi ft + j)S_{input} S_{input}$; and mixer $= \cos(2\pi ft + j)S_{input}$.

```
if(sample == 1)
      Idemod(t)   = ampl*Ss1(t)*mixerI/longsample;   %Ss1(t)*cos(2*pi*fc*gg);
      Qdemod(t)  = ampl*Ss1(t)*mixerQ/longsample;   %-Ss1(t)*sin(2*pi*fc*gg);
else
      Idemod(t) =   ampl*Ss1(t)*mixerI/longsample + Idemod(t-1);
      Qdemod(t) =   ampl*Ss1(t)*mixerQ/longsample + Qdemod(t-1);
end

  if((lock==0)&&(sample == longsample)&&(((((I(inc) == 1)&&(Idemod(t) >= .95))||
      ((I(inc) == -1)&&(Idemod(t) <= -.95)))&&(((Q(inc) == 1)&&(Qdemod(t) >= .95))||
      ((Q(inc) == -1)&&(Qdemod(t) <= -.95)))))
      lock = 1;
  end

  if((sample == longsample)&&(lock == 0))
      phasedemod=phasedemod + pi/64;
  end

  if(phasedemod >= 2*pi)
      phasedemod = 0;
  end
  pLL(t)=phasedemod;     % save the phase for plotting

      sample = sample + 1;
  if(sample > longsample)
      sample = 1;
      inc = inc + 1;
  end
```

Figure 1: Matlab implementation of the Demodulator.

Several functions, such as the mixers in the demodulator, will not need simulation through the whole cycle, because the program is communicating or in phase with the incoming signal close to 100%. The program is then allowed to reduce its frequency by half the cycle to receive the full value of an incoming signal. The phase angle depends on the PLL that is controlled by an enable switch (when the Sinput is in phase). The Matlab algorithm and VHDL codes give details on how the PLL was implemented. The Matlab model for the demodulator and PLL function are shown in Figure 1, and the frequency lowering algorithm is given in Figure 2. The PLL will force the frequency to be lowered based on when the output signals of the demodulator can reach its maximum value. This will cause the simulator to skip several variables when the demodulator reaches its value at a faster pace before the end of each period.

```
if((1 > longsam)&(cycle == cycletot)&(lock == 1)&(phasecount > corang))
    breakm = breakm*2; fprintf('max');
else if((1 > longsam)&(cycle == cycletot)&(lock == 1)&(phasecount <= corang1 ))
    fprintf('min');   breakm = breakm/2;
end
end

if(breakm < 1)
    breakm = 1;
else if(breakm > 4)
    breakm = 4;
        end

if(1 == 1)
    II(samp)    = freqmultadj*sqrt(2)*y2(samp)*mixerI/longsam;
    QQ(samp)   = freqmultadj*sqrt(2)*y2(samp)*mixerQ/longsam;
else
    II(samp) = freqmultadj*sqrt(2)*y2(samp)*mixerI/longsam+II(samp-1);
    QQ(samp) = freqmultadj*sqrt(2)*y2(samp)*mixerQ/longsam+QQ(samp-1);
end
```

Figure 2: The frequency lowering algorithm.

Table 1: PC instruction commands.

Data from PC	Commands
0001	ALU C= A – B
1001	ALU C= A + B
0010	ALU C= A AND B
0011	ALU C= A OR B
0101	ALU C= A
0110	ALU C= A mult B
0100	Jump back to Register value
0101	write to memory
1000	A=mem(B)
1001	mem(B) = A
1010	If(Z/=0),PC=value
1011	If(Z=0),PC=value
1100	PC=value
1101 0000	A = B
1101 0010	A greater than B

Table 2: Floating numbers

Significant bits	6	5	4	3	2	1	0
Decimals	1	.5	.25	.125	.0625	.03125	.0115625

Components such as adders/subtracters and multiplier divider were added to perform the math functions for the DSP chip. The ALU components, PC instruction, and floating numbers are given in Tables 1–3.

Table 3: Add or subtract input command.

Add or Subtract	Z = 1 if B > A	ALU Operate	most sign bit
A + B	Null	A + B	Pos
-A + -B	Null	A + B	Neg
A + -B	0	A − B	Pos
A + -B	1	B − A	Neg
-A + B	0	A − B	Neg
-A + B	1	B − A	Pos
A − B	0	A − B	Pos
A − B	1	B − A	Neg
A − -B	NULL	A + B	Pos
-A − B	NULL	A + B	Neg
-A − -B	0	A − B	Neg
-A − -B	1	B − A	Pos

DESIGN

With increasing pipelining by N sections, the voltage may be reduced by V/N. The total power will be reduced by N2. Quadratic reduction in power consumption is one advantage of increasing pipelining. Another advantage is overhead that is typically much less than that of parallelism. This means less hardware objects are needed for pipelining compared to parallelism. There are disadvantages when using pipelining. Not all algorithms or programs are recommended for pipelining. Adding a pipeline structure to a program usually increases the error in branching.

Figure 3 displays a block for a Xilinx board that is capable of handling small designs to perform the various functions. It can perform the same tasks compared to an ASIC design and can be developed at a quicker rate. This Xilinx board was used to prove that the waveforms in Matlab and VHDL are capable of simulating in real time. Xilinx uses a different format to synthesize VHDL codes compared to ASIC Implementation. There were RAM and ROM chips that contain the same data information as in VHDL codes that were used in the Xilinx dictionary.

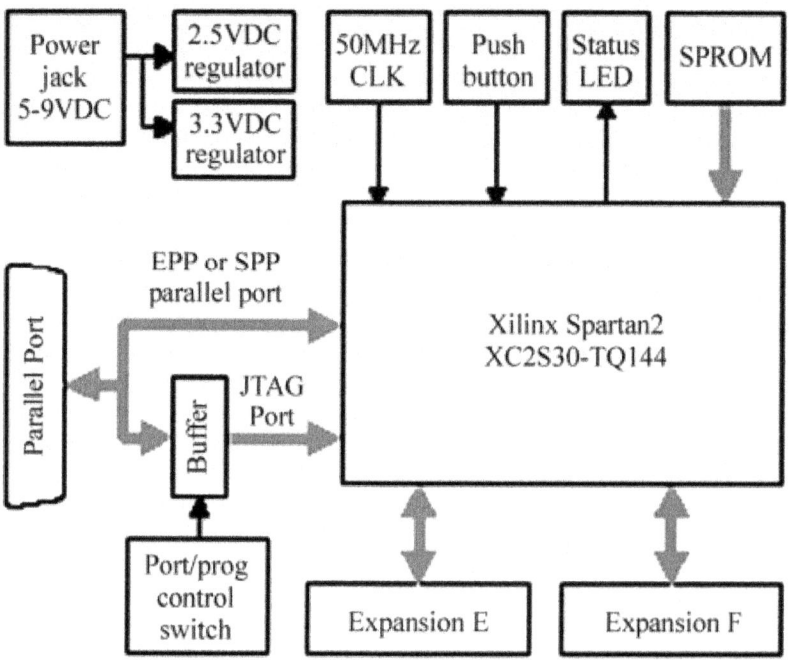

Figure 3: Xilinx FPGA board - XCQS30-TQ144.

This will save memory space inside the FPGA because of the familiarity format of the memory design. Figure 4 displays a command file to synthesize some of the same VHDL files that were synthesize in the ASIC design. To simulate this command file a user needs to type in the following command: fc2_shell –f cell16_xilinx. This command will also assign the input and output pins to the variable inside the main VHDL code. There were no errors after the last command. The next was to simulate the next command file in Figure 5. This file will create a bit file to be downloaded into the FPGA unit. If no errors have occurred then it will simulate the same way as it has simulated in Mentor Graphics.

```
create_project cell16_xilinx

add_file –library WORK-format VHDL cell16.vhd
add_file –library WORK-format VHDL dspmath.vhd
analyze_file –progress

create_chip –progress –name cell16 –target SPARTAN2 –device 2S3OTQ144
-speed -5 –frequency 50 –fast –preserve cell16
current_chip cell16

set_pad_buffer BUFGP /cell16/clk              set_pad_loc P91 /cell16/clk
set_pad_loc P93 /cell16/digitalout<7>         set_pad_loc P95 /cell16/digitalout<6>
set_pad_loc P99 /cell16/digitalout<5>         set_pad_loc P102 /cell16/digitalout<4>
set_pad_loc P112/cell16/digitalout<3>         set_pad_loc P114 /cell16/digitalout<2>
set_pad_loc P117 /cell16/digitalout<1>        set_pad_loc P120 /cell16/digitalout<0>
set_pad_loc P80 /cell16/reset                        set_pad_loc P84 /cell16/button
set_pad_loc P59 /cell16/digitalin<1>          set_pad_loc P57 /cell16/digitalin<0>

optimize_chip –name cell16-Optimized-progress

export_chip –dir.

list_message

exit
```

Figure 4: Command file for synthesizes.

```
#!/bin/sh
PATH=$PATH:/export/eda/Xilinx/ise5/bin/sol
export PATH
LD_LIBRARY_PATH=/export/eda/xilinx/ise5/bin/sol
export LD_LIBRARY_PATH
XILINX=/export/eda/Xilinx/ise5
export XILINX
XIL_MAP_LOCWARN=""""
Export XIL_MAP_LOCWARN

ngdbuild –p 2s30-5-tq144 cell16.edf
map –p 2s30-5-tq144 –o map.ncd cell16.ngd cell16.pcf
par –w –ol 2 –d 0 map.ncd cell16.ngd cell16.pcf
trce cell16.ngd.ncd cell16.pcf –e 3 –o cell16.twr
bitgen –g StartupClk:JtagClk –l –w cell16.ngd.ncd
ngd2vhdl –w cell16.ngd cell16_xilinx.vhd
```

Figure 5: Command file for FPGAs unit.

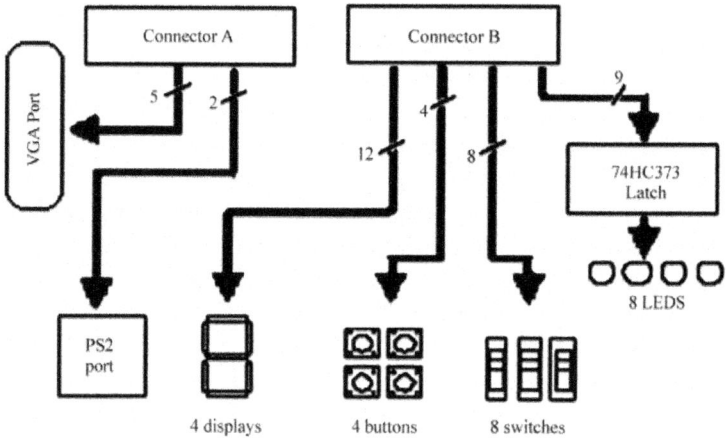

Figure 6: Xilinx input and output connect.

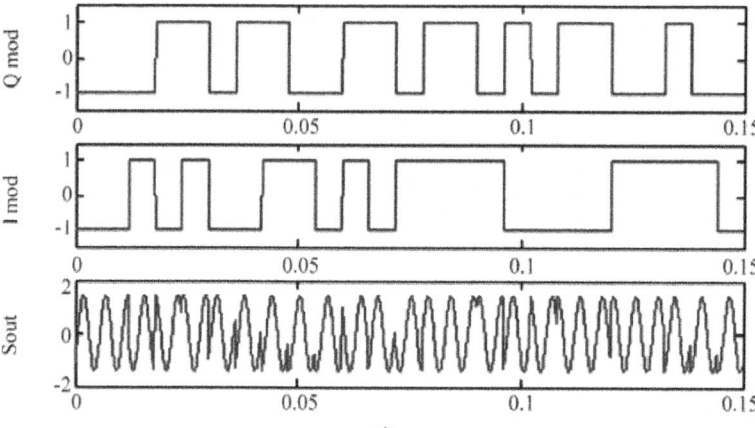

Figure 7: Transmitted output (Sout) with input (I mod, Q mod) at 200 Hz

Figure 6 displays components for 8 switches, 8 LEDs, 4 buttons, and a seven-segment display. For this design switches, LEDs, and buttons was able use to verify the electronic components works. Switch(8) was used to reset the FPGA microchip in the same way as it was going to be used in Mentor Graphics. Switch(1) and switch(2) were used to input data for the modulation function. They represented the signals digitalout (Qmod) and digitalout1(Imod) respectively. The 8 LEDs, button(1), and switch(3) were used to represent the output of the demodulation function. When button (1) is equal to zero(not pressed) it represents the first eight binary numbers of the output signal. When it is equal to one it represents the last eight binary numbers and LED(7) indicate if Iout or Qout is positive or negative. Switch(3) was used to indicate if the LEDs represent Iout or Qout. If switch(3) is equal to one then it will represent Iout, otherwise it will represent Qout. Button(2) was used to display the output results from the LEDs at the end of every period. All the input and output signals from the Xilinx board match with the input and output waveforms from Mentor Graphics, while being tested for verification. This proves that the microchip design would have worked if it were processed on an ASIC design chip.

RESULTS AND DISCUSSIONS

A modulator function was written in Matlab to support the transmitter component. The QPSK method was used to output the signal Sout(time). The Sout variable is an analog signal that communicates through any base station with other cell phones. Figures 7–9 display the inputs of the message codes with the output of the Sout transmitter with a frequency of 200Hz. Notice how the phase shifts in Sout(signal) when the inputs, Q(t) and I(t), change with respect to time. In Figure 10, the Q(demod) and I(demod) represent signals from the demodulator function or receiver component when all signals are in phase. They determine whether the value of the code messages is either a 1 or -1 by sloping to its maximum value before the clock period starts over. Figure 11 displays the same waveforms when they are out of phase. The Q(demod) and I(demod) do not reach their maximum value at the same time at the end of every clock cycle. When this occurs the receiver is not detecting any signal or the demodulation mixer is out of phase with the incoming single (Sout).

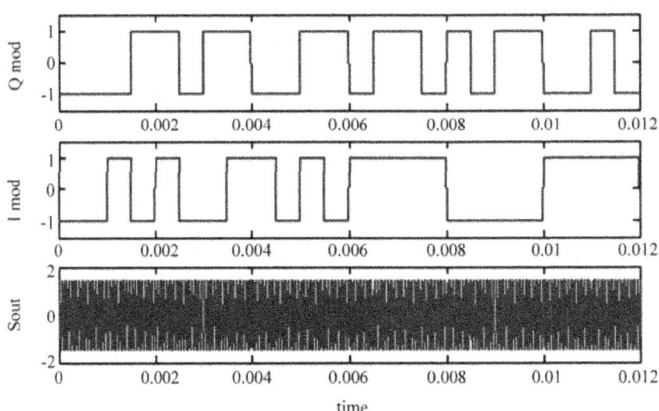

Figure 8: Transmitted output (Sout) with inputs (I mod, Q mod) at 20,000 Hz.

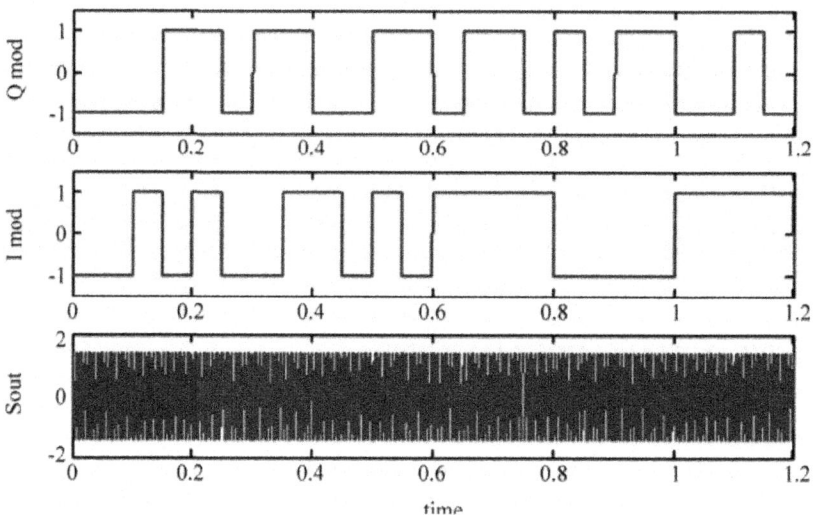

Figure 9: Transmitted output (Sout) with inputs (I mod, Q mod) at 2,000,000 Hz

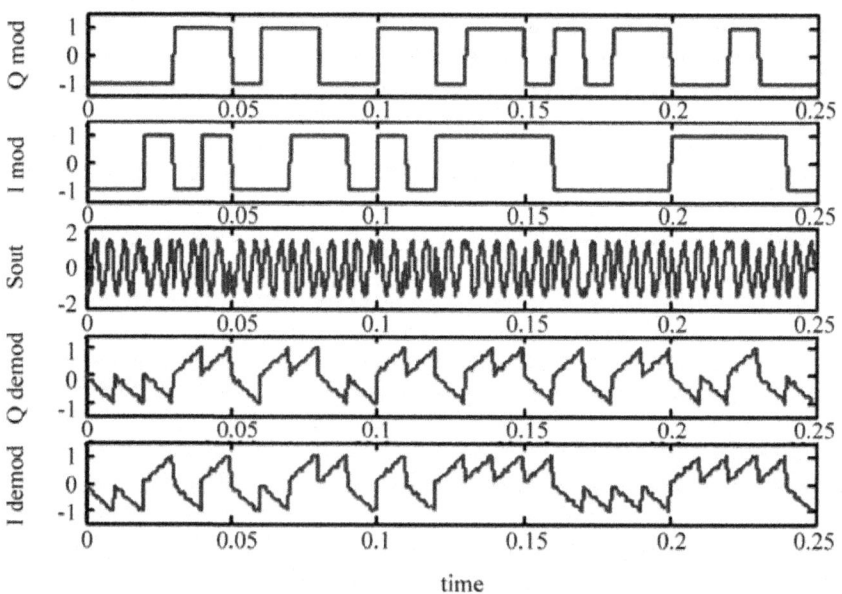

Figure 10: In phase demodulation function.

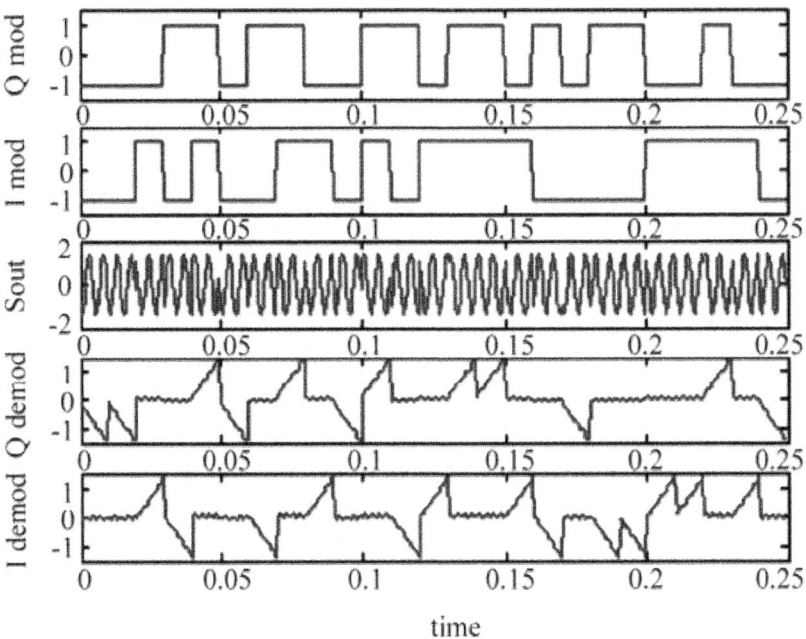

Figure 11: Out of phase demodulation function.

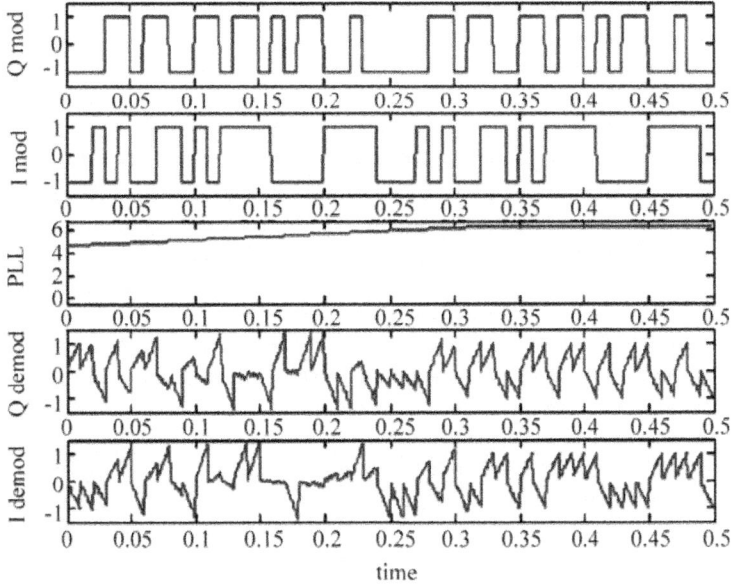

Figure 12: Searching the phase with a Phase Lock Loop.

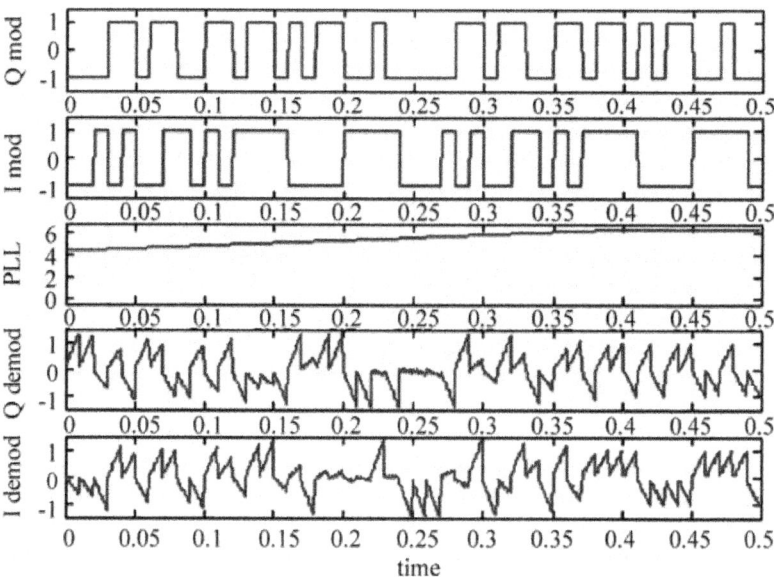

Figure 13: Searching for the exact phase with the Phase Lock Loop.

There are several ways to implement the PLL function inside the demodulator, depending on the communication software requirement. For this design the PLL is one of the main sources to reduce power in the cell phone system. Figure 12 displays waveforms to show that once the PLL is locked then the correct message symbol will be decoded. Several times the mixer from the demodulator can have the same phase with different amplitudes. Figure 13 displays that the Q(demod) and I(demod) reach their maximum value but one or both values have the wrong value as the message code (Imod, Qmod). Initially, the electronic component inside the cell phone needs to know what sequence of message code it is supposed to receive before it freezes the PLL.

The initial sequence message codes vary, depending on the communication network and the base station that will be described in the next section. Several filter functions was able to be tested to filter out noise for modulation signals. Programs to generate noise, such as white noise and fading noise (Rayleigh and Rician noises), were written to create noise for output modulated signals. The programs for filter functions were proved effective, allowing that multiple cell phone users to still communicate through the noise disturbance in the incoming signal. There were several algorithms that were developed to handle the type of noise to minimize the error in the message codes. Figure 14 displays a transmitter signal with two types of noise.

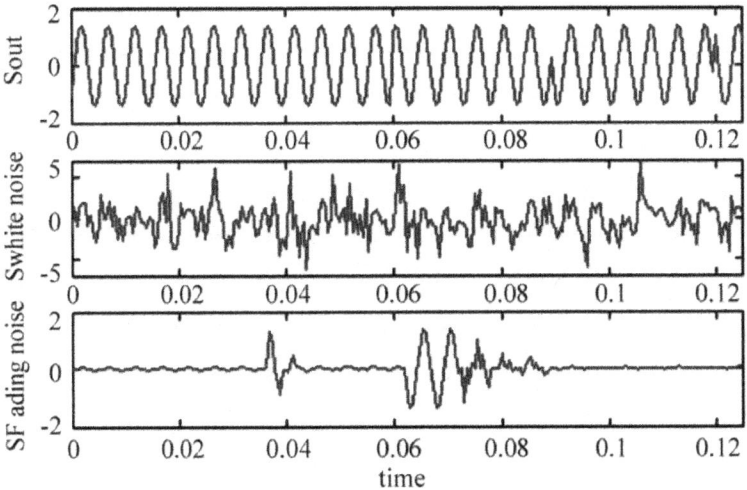

Figure 14: Noise waveforms.

A band pass filter technique (Butterworth and Chebyshev) was used to filter out white noise. The signal to noise ratio is a major factor to filter out the unwanted magnitude of the transmitted signal. Based on simulation comparisons, if the noise amplitude is more than half the size of the transmitted signal then there is a higher probability that an error will occur compared to a lesser amplitude noise signal. Figure 15 compares both of the filter functions that were simulated in the compiler and in the program algorithm to filter out the two frequencies that were out of range for the band pass filter. The outputs from both of the filter contain the same results after comparing them. Figures 16 through 19 displays several filter results that vary based on the maximum amplitude of white noise.

Notice that the lower SNR decrease the higher probability that the receiver will not decode the message's incoming signals. Figure 20 shows that the message code from CDMA that is capable of being demodulated. (The Matlab implementation for this is given in Figure 24). Figure 21 shows lowering the frequency. Figure 22 gives the tested waveforms as simulated with Mentor tools. Another test sample was used to test the modulation and demodulation process when they were initially out of phase. This will cause the signal values of digitalout and digitalout1 to not reach the maximum value of 1 or -1 at the end of each period. In Figure 23 the maximum value of the output at the end of each period have increased or decreased because it tried to find the exact phase for the Phase Lock Loop.

CONCLUSIONS

The tested waveforms that were viewed in Mentor Graphics matches with all the waveforms in Matlab. This means that the microprocessor is capable of performing modulation and demodulation techniques with other communications systems. Also, since the LEDs outputs from Xilinx matches with the waveforms in Mentor Graphics indicate that the Xilinx can be implemented as cell phone microcontroller or a base station controller.

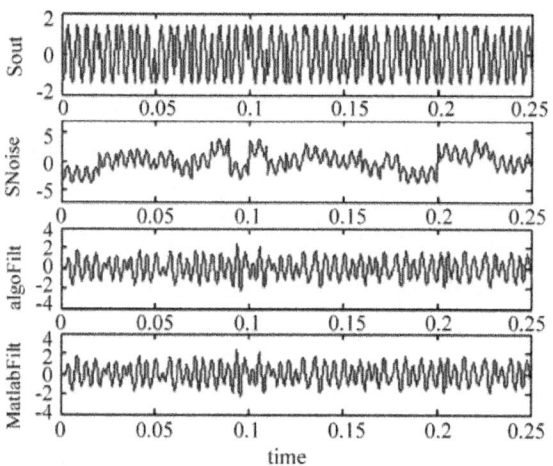

Figure 15: Comparison of filters' performance.

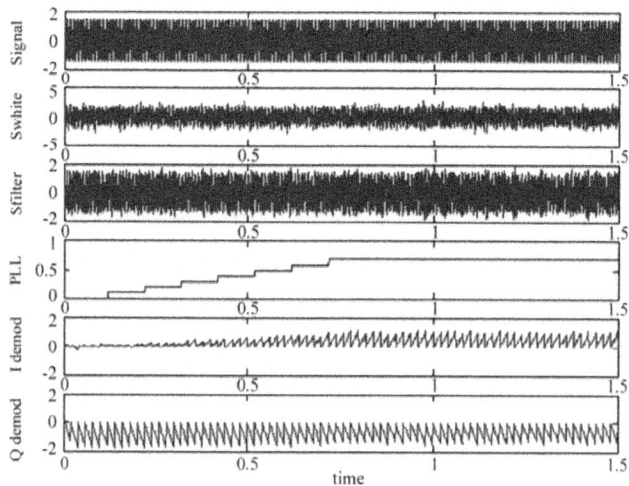

Figure 16: White noise 25% of message signal.

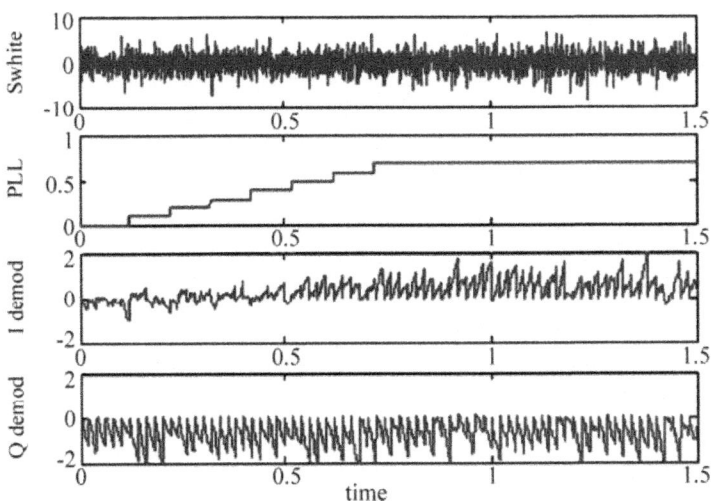

Figure 17: White noise 100% of message signal.

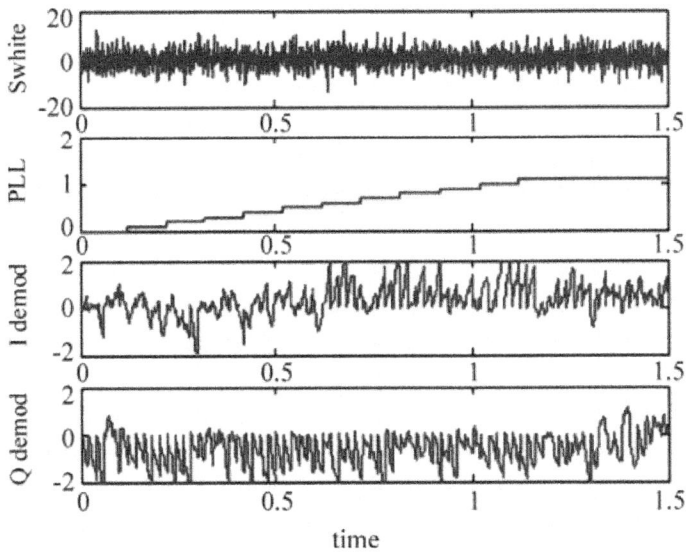

Figure 18: White noise 200% of message signal.

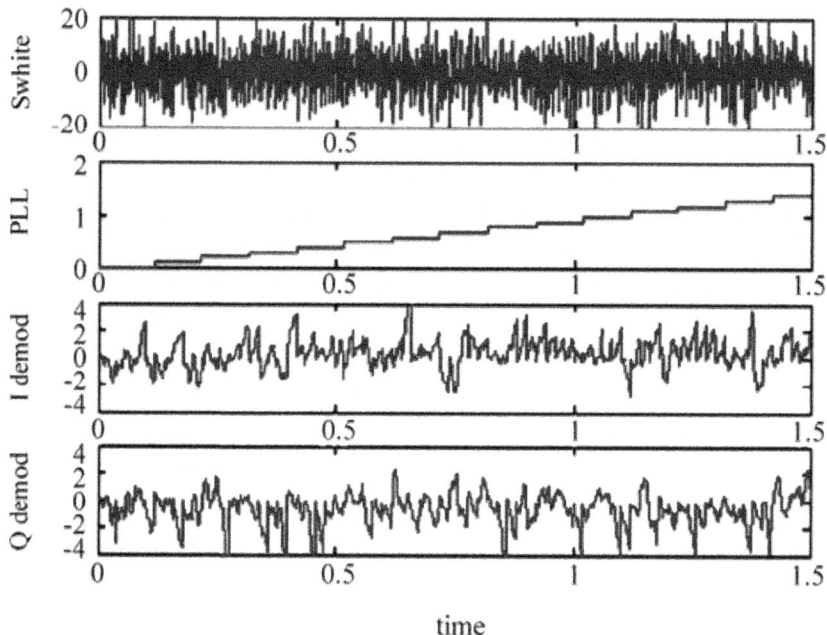

Figure 19: White noise 400% of message signal.

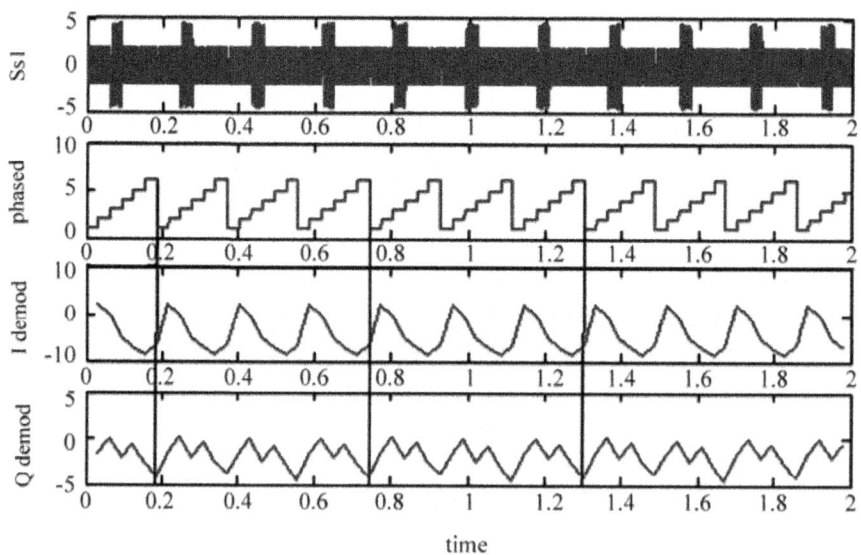

Figure 20: CDMA process in Matlab.

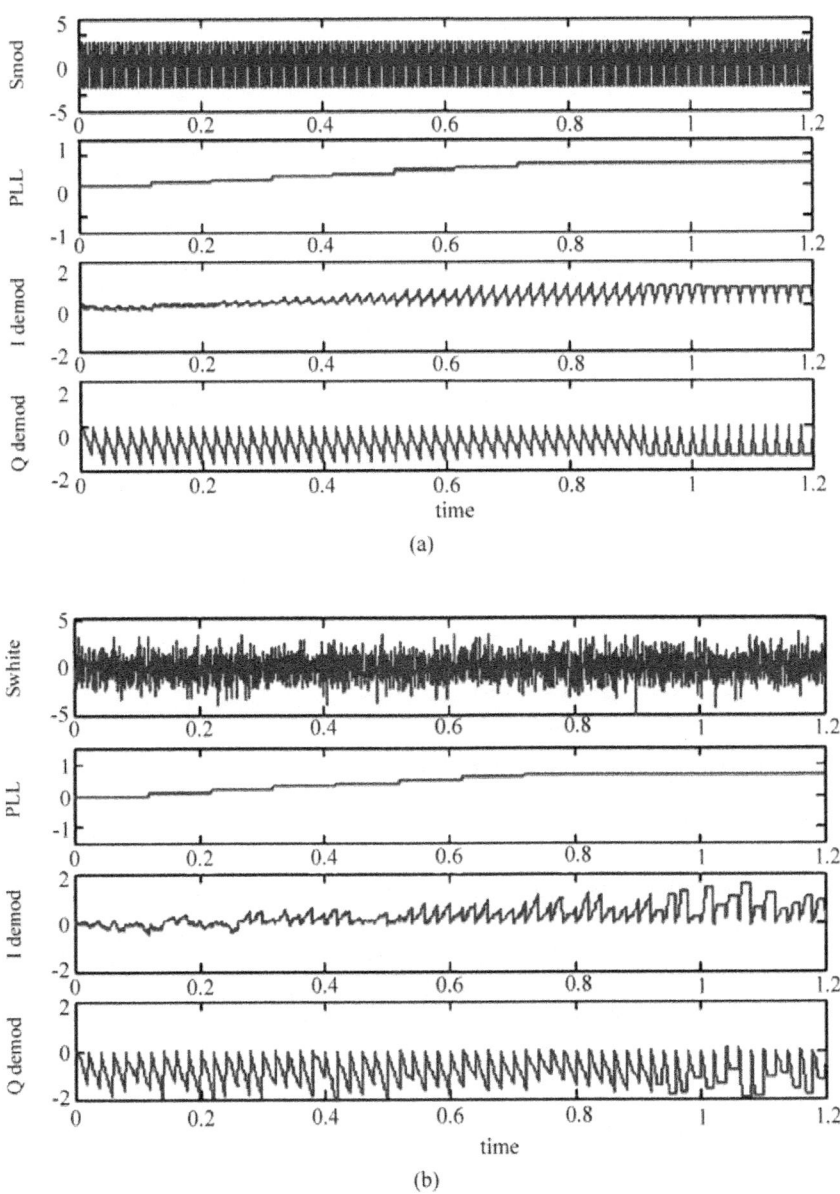

Figure 21: (a) Lower the frequency; (b) Lower the frequency with noise.

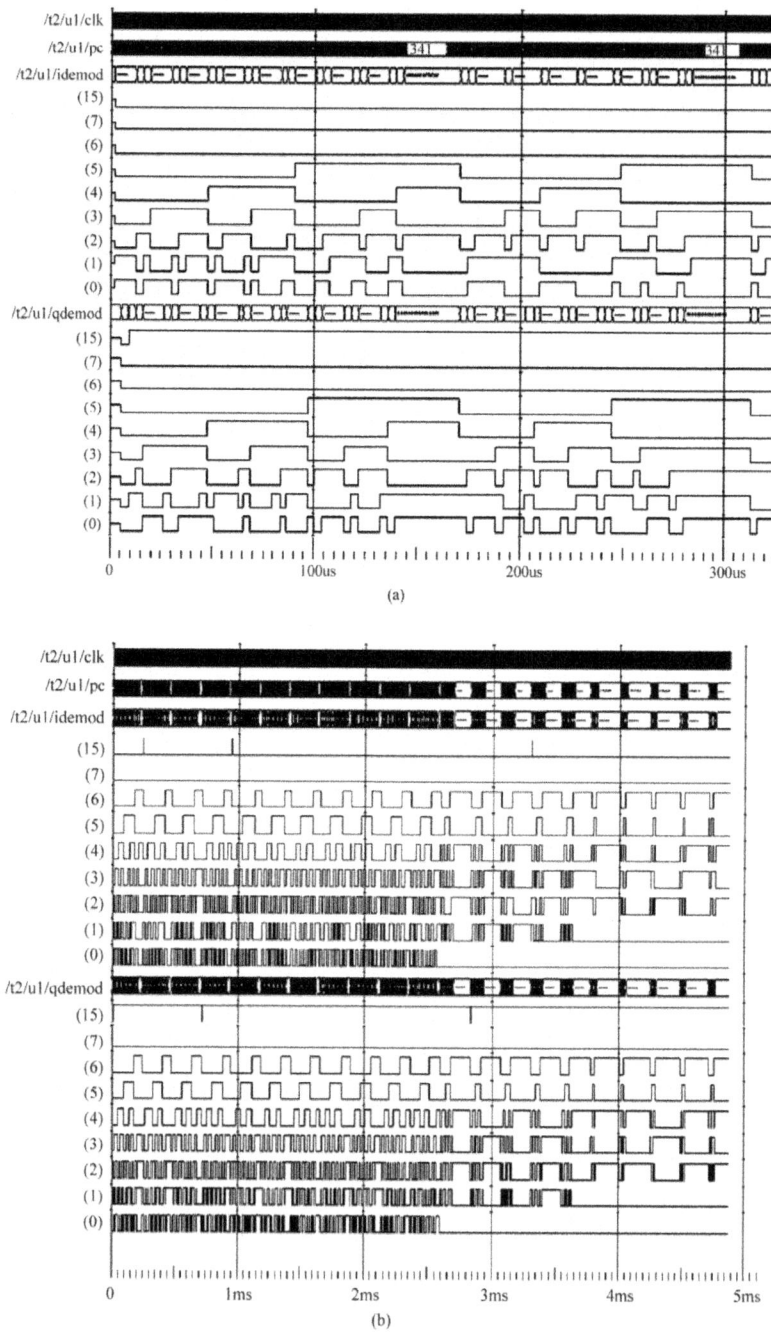

Figure 22: (a) Tested waveforms for the first two periods; (b) Tested waveform while lowering the frequency.

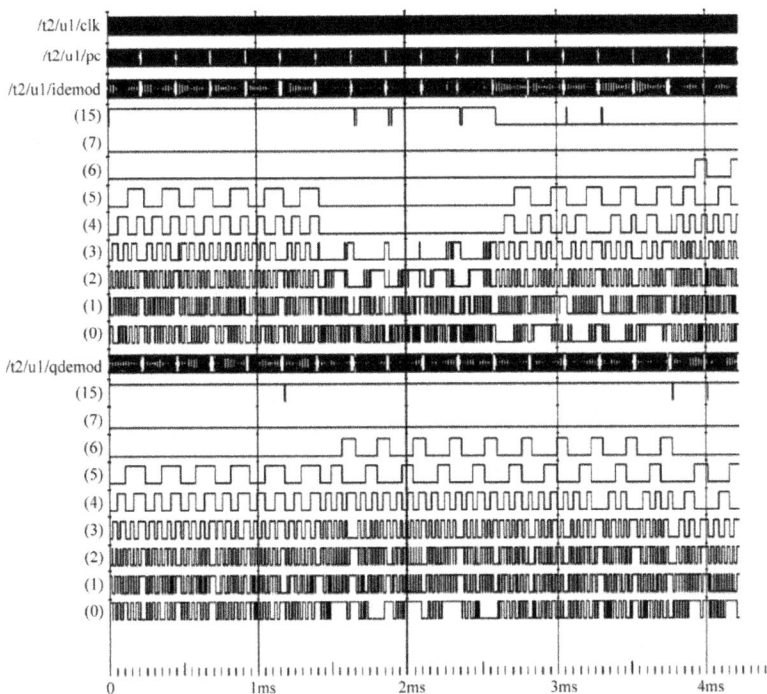

Figure 23: Tested waveforms while out of phase.

```
Icdma(samp) =   input1a*seqnum(sq),          Qcdma(samp) =   input1b*seqnum(sq);
 Icdma2(samp) = input2a*seqnum2(sq);         Qcdma2(samp) = input2b*seqnum2(sq);
 Icdma3(samp) = input3a*seqnum3(sq);         Qcdma3(samp) = input3b*seqnum3(sq);
 Icdma4(samp) = input4a*seqnum4(sq);         Qcdma4(samp) = input4b*seqnum4(sq);


Icdmat(samp) = Icdma(samp) + Icdma2(samp) + Icdma3(samp) + Icdma4(samp);
Qcdmat(samp) = Qcdma(samp) + Qcdma2(samp) + Qcdma3(samp) + Qcdma4(samp);
Ss1(samp)    = Icdmat(samp)*cos(2*pi*freq*t+ phasein) + Qcdmat(samp)*sin(2*pi*freq*t+ phasein);

     if(1 > longsam)
        times1(samp1) = t;
          if(sq ~= 1)
             IIz(samp1) = II(samp)*seqnum2(sq)+IIz(samp1-1);
             QQz(samp1) = QQ(samp)*seqnum2(sq)+QQz(samp1-1);
          else
             IIz(samp1) = II(samp)*seqnum2(sq);
             QQz(samp1) = QQ(samp)*seqnum2(sq);
          end
          if(sq == 6)
             IIcdma(samp2)=IIz(samp1);      QQcdma(samp2)=QQz(samp1);
             times2(samp2)=t;                 samp2=samp2+1;
          end
     end
```

Figure 24: Matlab implementation of CDMA.

The methods were the basic functions that all cellular phones use for communication with the base station. There were ideas that were created to adjust several algorithm functions inside the DSP chip to save power. Lowering the frequency is the main method that was developed to save power in DSP chips compared to pipelining, parallelism, and reducing several algorithms for several functions such as the cos(x) and sin(x). There are several adjustments that might not have been recognized to simulate with the functions inside the main microchip controller. This study proves that these functions work in three different real time compilers, which are Matlab, Mentor Graphics, and Xilinx. This proves that these methods can be developed in any type of microchip fabrication factory. The Mach-Pa compiler in Mentor Graphics proves that if the frequency was lowered, it would automatically save power compared to when it is higher. This has been an interesting and challenging project to promote a better understanding of how the communication process worked with the base station and cell phones. Other components to have a better internal knowledge of working with the main microchip are the push buttons, screen display, microphone, speaker, and antenna for every cell phone.

REFERENCES

1.   Montanaro, R. T. Witek, K. Anne, A. J. Black, E. M. Cooper, D. W. Dobberpuhl, P. M. Donahue, J. Eno, W. Hoeppner, D. Kruckemyer, T. H. Lee, P. C. M. Lin, L. Madden, D. Murray, M. H. Pearce, S. Santhanam, K. J. Snyder, R. Stehpany, and S. C. Thierauf, "A 160-MHz, 32-b, 0.5-W CMOS RISC microprocessor," IEEE Journal of Solid-State Circuits, Vol. 31, No. 11, pp. 1703–1714, November 1996.

2.   F. M. Gardner, "Phaselock techniques," 2nd Edition, John Wiley and Sons, New York, NY, 1979.

3.   R. Philips, "Low power PLL FM demodulator for satellite TV receivers, integrated circuits," Trenton, NJ, 1996.

4.   M. Vesterbacka, "Digit-serial implementation of LDI/ LDD allpass filters," Scottsdale, AZ, 2001.

5.   H. Ming, O. Vainio, and M. Renfors, "Digit-serial design of a wave digital filter," Proceedings of IEEE Instrumentation and Measurement Technology Conference, Vol. 1, pp. 542–545, 1999.

6.   Fettweis, H. Levin, and A. Sedlmeyer, "Wave digital lattice filters," International Journal of Circuit Theory and Applications, Vol. 2, pp. 203–211, June 1974.

Chapter 3

USES OF ELECTRONIC COMMUNICATION TO DOCUMENT AN ACADEMIC COMMUNITY: A RESEARCH REPORT

Anne J. Gilliland-Swetland and Greg Kinney

Council on Library and Information Resources 1755 Massachusetts Avenue, NW, Suite 500 Washington, DC 20036

ABSTRACT

The Bentley Historical Library Computer Conferencing Appraisal Project (NHPRC Grant No. 9 1 - 1 13) began in September 199 1. Its aims were to explore whether or not computer conferencing has potential to document the intellectual, cultural, and social environment of colleges and universities, and then to make recommendations regarding the archival appraisal and accessioning of such materials. Two larger purposes of the project were 1) to propel the Bentley Library into facing the archival challenges-both intellectual and practical-presented by electronic communication at the University of Michigan and 2) to raise the level of discourse in the college and university archival community about the nature and administration of materials generated through electronic communication. This article provides an overview of computer conferencing systems; in particular the CONFER I1 software used at the University as well as many other academic and non-academic institutions, and then outlines the methodology and major findings of the project. The article also discusses some issues raised by the project that may well be generic to the archival management of electronic communication.

INTRODUCTION

Helen Samuels has argued that the prime responsibility of academic archives is to document the functions of colleges and universities. These functions are: conferring credentials, conveying knowledge, advancing knowledge, maintaining culture, providing public service, socializing students, and sustaining the institution.' College and university archivists with a mission to document their academic environments, however, have long realized that, while the administrative aspects of their institutions tend to be well-documented through organizational records, intellectual discourse, pedagogy, and student culture are not.' To translate this realization into Samuels's terms, college and university archivists have not been very successful in documenting the environment and processes in and by which academic institutions convey and advance knowledge, maintain culture, and socialize students.

There are a variety of reasons for the difficulties inherent in documenting these functions. Almost no administrative records are created that capture the discussions and interactions underlying the genesis and transfer of ideas and opinions in the many disciplines, professions, and ways of life present on the typical college campus. The personal papers of faculty sometimes give limited insight into these areas, through correspondence, publications, and research notes. These insights, however, seem to have declined steadily during the latter stages of the twentieth century, in large part due to the increased use of the telephone, a communication medium that leaves no documentary trail.

With the obvious exception of curricular development discussions, teacher-student interaction, in terms of both how material is presented to students and the mentoring role assumed by faculty, is also poorly documented in administrative records. Faculty personal papers or student collections may prove useful for such study because of the occasional availability of lecture notes and other course material; even with these sources, however, much information of potential value to intellectual and cultural historians and researchers in the social sciences is not recorded. Student life is another area that traditional archival records and personal papers do not document well. While records documenting various aspects of student life are available in narrowly focused areas, the overall intellectual and social experience is difficult to reconstruct. Commonly available in the archives are such official university documents as academic counseling material or housing office records, which, together with records of student organizations, student publications, football programmes, fraternity and sorority material, ephemera such as flyers and posters, and similar documentation, do provide a picture of student life on campus. Unfortunately, it is a distorted picture, as such records represent only partially the student experience.

Archival staff at the Bentley Historical Library were intrigued by the possibility that the growing academic use of electronic communication technologies might result in additional documentary sources that could partially address the shortcomings found in traditional archival material. With funding from the National Historical Publications and Records Commission

(NHPRC), the Bentley undertook a one-year appraisal project to look more closely at the documentary potential of electronic communications. The project staff decided to focus upon computer conferencing because it is an electronic communication medium that has been in use for a comparatively long time (from a technnlogical perspective), and many early text files still exist. Conferencing appeared to have a particular appeal to those in an academic environment and enjoys widespread use, not only at the University of Michigan but in colleges and universities throughout North America and Europe.'

A review of the research literature and methodologies of several disciplines indicated that computer conferences might provide a significant research resource for social scientists as well as historians. Indeed, a considerable amount of research has already been conducted into electronic conferencing since its inception in the early 1970s. Particular focuses of this research include the establishment of communication norms and investigation of new group dynamics, content and lexical analysis, the use of conferencing in computer-mediated interactive instruction, and the "invisible college" value of computer conferencing as a mechanism of scholarly communication.

There were also wider professional reasons driving this project, most notably the pressing need for college and university archivists to begin to address the intellectual and practical challenges presented by electronic communication. The administrative importance of electronic record-keeping applications had long been a focus of concern and research among governmental archivists. The extent to which the approaches and practices developed by government archivists were transferrable to different organizational environments such as those of colleges and universities, and not necessarily viewed within the strict parameters of record-keeping electronic media, remained to be tested. Pennsylvania State University Archives has investigated the applicability of existing archival approaches to university administrative data file The Bentley Historical Library elected to explore computer-mediated communication-specifically, computer conferencing-in an environment with multiple facets extending beyond the administrative.

OVERVIEW OF COMPUTER CONFERENCING AND THE DEVELOPMENT OF CONFER 11 CONFERENCING SOFT-WARE

Before discussing the methodology and specific findings of this project, it is necessary to provide some background both of the evolution and nature of computer conferencing as a whole, and of the implementation of computer conferencing at the University of Michigan. Indeed, a major premise underlying this project was that most archivists know little, if anything, about computer conferencing, despite the fact that it has been a widely utilized communications medium for over twenty years.

The development of computer (or electronic) conferencing directly reflects the rise of computer-mediated communication (CMC), or computer-based message systems (CBMS), in the 1960s and 1970s and their subsequent proliferation with the advent of the microcomputer era in the late 1970s. Today conferencing is widespread throughout higher education, scientific and research institutions, and the corporate se~tor.~ Ellen Pearson has provided a succinct definition of a computer conference as: an ongoing database of all text contributed by the conference members. Members may search for and retrieve stored text at any time. Typically, participation in the conference is asynchronous .... Each participant sees all the others' statements and may comment on those already entered andlor add new thoughts to the discussions. The conferencing system software tracks all new entries, linking statements and comments thereto, so that members may read or proceed through the messages either chronologically or logically. The software also tracks each member's individual online session so that when he next joins or signs on, he is notified of the numbers of new or unread messages.'

Conferencing systems all share the basic characteristics laid out by Pearson, although the style and "feel" of conferences can differ considerably from one system to another. Some systems favor short-term, small group projects, while others are designed to foster extensive, ongoing dialogue in a wider setting, and utilize more sophisticated text entry and organizational capabilities. Three of the earliest and most influential systems were EIES (earlier versions of which were Partyline, Discussion, and EMISARI), PLANET (originally FORUM), and CONFER. Other common conferencing systems include DEC's VAX NOTES and Honeywell Multics' FORUM. EIES (Electronic Information Exchange System) was sponsored by the National Science Foundation (NSF) and developed by Murray Turoff at the New Jersey Institute of Technology. PLANET developed out of research led by Jacques Vallee and Hubert Lipinski at the Institute for the Future in Menlo Park, California. One of the primary institutions associated with PLANET was the University of Southern California in Los Angeles. CONFER was developed at the University of Michigan by Robert Parnes and first implemented in 1975 just at a point when work in group communications media was beginning.x Parnes developed the first version of the CONFER software for his doctoral dissertation in Educational Psychology through the University's Center for Research on Learning and Teaching (CRLT) and with partial support from NSF funds designated for development of electronic communication in support of academic activity. The development of the CONFER software is documented at the Bentley in the files of CRLT as well as in Parnes's doctoral dissertation."

Parnes's aim with CONFER was to design a system to: address both the communication and decision making needs of university and other groups involved in governance ...[and to] facilitate small group communication ... limited to less than a thousand people .... The set of principles which I have tried to operationalize in the CONFER system are those of individual equality, freedom, privacy and flexibility, and the facilitation of individual participation.

Other principles upon which Parnes based his design were really quite visionary for the mid-1970s. They included the assumptions that the primary users would be geographically widely dispersed, computer novices, and untrained in interactive computing; that eventually most if not all users would have their own personal computer or computer terminal; and that users would be making a long-term commitment to computer conferencing by using it on a day-to-day basis for more than just task-oriented activities." Upon completing his dissertation in 198 1, Parnes set up his own company in Ann Arbor, Advertel Communication Systems, which develops and markets CONFER commercially. The rights to the software belong to Parnes with the university having non-exclusive rights to use his CONFER I1 software. Today CONFER I1 is one of the leading and most influential electronic conferencing software packages available.' It has been used by many colleges and universities throughout the United States, Canada, and Great Britain. In Michigan alone, these include Michigan State University, Wayne State University, and Western Michigan University. CONFER has also been implemented by other institutions such as the Research Libraries Group (RLG), the Bureau of Social Science Research, the University of Michigan Highway Safety Research Institute, Formative Evaluation Research Associates, Onyx, and Acumenics (as consultants to the Federal Aviation Authority).

At the University of Michigan between 1975 and 1991, over 3,100 conferences of many different types were hosted on the university's computers with over 165,000 individual memberships. Log-ins to conferences currently run at more than one per minute.I4 More than 250 conferences, both current and no longer active, have so far been identified on the union list under development by the Bentley Historical Library, a testament to both their permanence and their transcience. In his I98 1 dissertation, Parnes categorized conferences into several types based on intended function. These categories are a useful way of illustrating the variety of tasks for which conferencing was envisaged: task force, workshop, committee, special interest group, commission, assembly, congress, and general interest group. Each of these functional categories could be further refined with any combination of the following variables: well- or ill-defined membership; focused or full-scope content; and fixed or indefinite duration.

For the purposes of this project, however, the Bentley archivists found it more useful to classify University of Michigan conferences, in line with local use conventions, as either public or private. Public conferences are those open to participation by anyone with a valid University of Michigan computing account.

Private conferences have membership limited by some defined criteria. Within that classification, the project archivists defined three types of private conference: administrative, course-related, and social. Public conferences are usually thematic, e.g., women's issues or student government, but ma covers a wide range of topics. Administrative conferences may operate on a staining or an ad hoc basis. They are used as electronic meeting forums, for decision-making, or to circulate documents to specific groups. Examples would be the university deans' conference or search committee conferences. Course-related conferences are associated with a specific course and membership is restricted to the students in that course. They may be used to follow up on class discussions, to generate student writing, as part of mandatory class participation, or simply to hand out assignments. The content of some are graded, whereas others are used merely to facilitate communication between student and instructor. Private social conferences are organized by two or more persons with common interests. Membership is recruited through a variety of formal and informal mechanisms, sometimes with participants voting on whether or not to admit a candidate for membership. These private conferences are not listed or documented by the University Computing Center and read-access is restricted to members only.

Although by far the most popular and widely used, CONFER I1 does not have a monopoly on conferencing at the University of Michigan. There are several similar commercial and non-commercial-as well as academic-conferencing systems, such as USENET, available within the Ann Arbor area. One relative newcomer is GREX, implemented in 199 1 for non-academic, non-commercial users. The University of Michigan itself also maintains a Forum conferencing system, which was originally used predominantly by Computing Center staff; its primary users now are from within the College of Engineering. Each of the systems has its advantages and loyal partisans. The widespread acceptance of computer conferencing in the University of Michigan community suggests that Parnes's expansive vision was correct; conferencing as a form of computer-mediated communication has indeed taken on a life of its own over and above any specific task-oriented function, both at

Michigan and everywhere it is in use. It,is this aspect in particular that caught the attention of archival staff at the Bentley and led them to hypothesize that conferences might be generating and recording material of potential archival value.

METHODOLOGY

Because of the relatively small amount of archival research that has been conducted to date on different forms of electronic communication, there were no tested models for the project archivists to apply in toto for this computer conferencing project. For this reason, they devised their own methodological approaches based in part on those advocated in archival literature and in part on their own experiences from working in the academic environment.

The main component of this project was appraisal of electronic conferences: first as a record genre and then as individual record series. (At this point, at least, individual conferences, which may include multiple "volumes," are being treated as discrete record series.) The project archivists were seeking to determine whether electronic conferences could provide documentation of the college and university functions outlined by Samuels and defined in the mission statement and collecting policy of the Bentley Library's University Archives and Records Programme. It was originally hypothesized that, if ten per cent of the examined conferences were found to contain significant evidential or informational value, the university archives would consider electronic conferences a record format that should be examined and appraised at the series level. If that hypothesis proved correct, then individual conferences were to be appraised and compared with the existing no electronic holdings of the Bentley. For this comparison archivists would look at subjects, individuals, and dates covered, nature and depth of material, and textual extent, to determine if conferences contained material not covered elsewhere in the collections, or which significantly overlapped or enhanced existing collections.

The appraisal methodology, therefore, was formed within this framework. For political and logistical reasons the university does not maintain metadata on conferences, such as a comprehensive listing or index of existing conferences, their organizers, purpose, or subject content. The Bentley's archivists, therefore, spent a considerable amount of time throughout this project trying to locate and gain access to on fervencies. 'They used five different sources to compile a union list and identify conferences for appraisal: a short list of major public conferences selected, described, and publicized online and in publications by the University Computing Center; references to other public conferences made by participants of major conferences; references made in university publications such as departmental newsletters; personal discussions with individuals organizing conferences; and ongoing contact with Robert Parnes, who remains responsible for initializing all new conferences. It was through Parnes that project archivists were able to make contact with organizers of course-related conferences.

Course conferences were the first and most procedurally complex area tackled, not only because of privacy considerations but also because of the fact that the project started right at the beginning of a new school year. It is at this time that conferences for semester-long courses are established. Parnes receives electronic mail messages from instructors wishing to establish a course conference and giving him a computer account ID under which to do so. He then initializes each conference and sends a message back to the instructor (or that person's computer ID-he often does not know his or her actual identity) to indicate that the conference is ready to go. At the end of the semester, the conference is automatically terminated and purged by the Computing Center. There is, therefore, minimal contact between Parnes and conference organizers. Parnes met with the project archivists and agreed to attach a message to his reply to all conference organizers informing them about the Bentley conferencing project and indicating that he had passed their electronic mail ID on to the archivists so that they could contact organizers directly about the possibility of observing their conferences. The archivists then sent a form letter via e-mail to all organizers explaining the purpose of the Bentley project and asking for permission to observe the course conferences while they were active and potentially to accession any that might prove to have considerable documentary value. While a few organizers were only too happy to work with the Bentley, the rest ranged from reluctant to downright hostile, despite reassurances of, and a number of measures taken to ensure, the privacy of conference participants.

The reactions of course conference organizers had been anticipated by the project archivists, who themselves were very concerned with the possible legal implications of appraising and, more importantly, possibly accessioning course-related conferences. They were particularly concerned, in the absence of clear legal guidance, about the extent to which the Furnib Educational Rights und Privacy Act (FERPA) might come into play, as well as with establishing ownership of the textual content of the conferences. Consequently, they devised a model two-part student release form to be used with active course conferences, one part dealing with FERPA, the other with literary rights. Project archivists asked cooperating organizers to give these releases to their students to sign at the beginning of the semester and then to forward them to the project office. This procedure also signified the agreement of both the organizer and the participants that the project archivists would be permitted to observe these private conferences for the purposes of appraisal and possible accessioning. Three classes agreed to be observed and signed and returned release forms, although in one class there was one student who refused to sign a form. This did not become a problem since the archival appraisal found the conference to have no long-term value. Potentially, however, such a situation could be an issue if another conference were deemed archaically valuable, accessioned, and made available to researchers.

In total, the project archivists identified for appraisal fifty-four active and twenty now defunct conferences and sub conferences to which they were able to gain access physically within the time constraints of the project. Readers should bear in mind that this figure represents probably less than two per cent of the total number of conferences ever hosted on a University of Michigan system, and is also a skewed sample given that it was to a large extent self-selecting. It is hard, however, to envisage any way, given the intellectual freedom constraints of an academic environment, to draw a more representative sample for such a project.

Appraisal of the conferences comprised several activities:

- An assessment of the statements of sponsorship and purpose, indexes, item descriptors, span dates, and participant lists all contained in individual conferences, as well as the life cycle of individual items (most of this was conducted online);

- The creation of random lists of sample items using Minitab to randomize item numbers;

- Purposive sampling of discussion items using item descriptors to identify topics of known interest to the Bentley and "fat file" topics (i.e., a large number of responses to a given item); and

- Assignment of Library of Congress Subject Headings to appraised conferences to facilitate identification and comparison of their content and date coverage with those of the Bentley's existing extensive university collections that are described in RLIN using the MARC AMC Format.

Based on the findings of this appraisal component, the project archivists were able to go on to make recommendations regarding possible accessioning and future work that needs to be done to make computer conferences intellectually and physically accessible.

News of the appraisal project quickly spread through the conferencing community, on campus and beyond, with the result that the project archivists engaged in numerous lengthy discussions about the project and its possible implications.

These discussions took place as items in several conferences, through electronic mail, and in-person. In large part this was because the appraisal process quickly revealed the overwhelming concern of all involved about privacy and ownership of conferences. Some of this concern is inherent in computer-mediated communication and the genre of conferencing in particular, and some relates directly back to the values CONFER developer Parnes hoped were reflected in the structure of the software (i.e, individual equality, freedom, privacy, flexibility, and the facilitation of individual participation).' Parnes appears to have been extremely successful in achieving his goals: conferences have become largely self-moderating, with many participants strongly aware of the archival, historical, and privacy implications of this medium of communication. Conferencing encourages an atmosphere of democracy that leads many participants to believe they are co-owners of the forum.

The fact that conferences at Michigan are accessible to anyone with a user ID and a terminal, by name, pseudonymously, or anonymously, also complicates the issues of literary and institutional ownership. These factors raise obvious issues for archivists of how to determine provenance ownership, and responsibility for the long-term preservation of conferences as well as how to authenticate the textual content of conference discussions. Archivists have also found that when they attempt to negotiate these issues directly with conference participants, they run the danger of having a "chilling effect" on the conferences that is, changing the nature of the conference environment, its discussions, and, by implication, the historical record. Participants have stated that, while they are very aware that computer conferences may contain material of future research value, they also have reservations about archival preservation. These reservations arise from concerns about privacy, intellectual freedom, and the legal status of public and private conferencing within a public university.

Indeed many participants feel that conferencing resembles a telephone call or intimate conversation among a group of friends. They are only too aware that CONFER was deliberately structured, both philosophically and technically, toward protecting group and individual privacy, and many have stated that potential "archiving" flies in the face of the private atmosphere fostered by CONFER. Participants too are concerned that as a result of this issue, the possibility of accessioning by the archives is changing the nature of the conference by inhibiting discussion or the use of personal names. The archivists have found that if they do not respond to these concerns, participants can effectively prevent conferences from being accessioned by the archives in a number of ways. Consequently, the project archivists have devoted a large amount of time to allaying participants' fears about potential "archiving."

The concerns about privacy are also institutionalized in the form of the University of Michigan's Conditions of Use of the Resources of the Information Technology Division Statement, which all users are required to sign or acknowledge before gaining access to any system. The policy states several conditions to which users must adhere. One of these is that the user agree in advance "to respect the privacy of other users; for example, you shall not intentionally seek information on, obtain copies of, or modify files, tapes, passwords belonging to other users or the University, or represents others, unless explicitly authorized to do so by those users." It is unclear what weight such an agreement might have in a court of law, but it does present a problem to the archivists if they are to accession and make available computer conferences on the grounds that they are "public records."

On the question of provenance, Pames writes that: The conference organizer derives herhis authority from the person who set up the conference and pays for the computer disk resources that it takes to support the conference. That is, there is a cost to maintaining the conference information on disk, and there is one person who is ultimately responsible for seeing to it that the computing system is properly reimbursed for providing the disk resources. That person is able to designate a conference organizer. Often the person simply assumes the role her himself. Thus the organizer is the owner (or officially speaks for the owner) of the conference files. By providing differential access to various parts of the conference files, CONFER is able to extend a kind of joint ownership to the author of each item of vote but it is the owner of the files who has ultimate control (and responsibility) over their contenk.

One result of these discussions and Parnes's views on ownership is that project archivists have down-played the question of accessioning conferences under authority of public record law. Concern over the issue is being met in part through a notice that is now being displayed as part of the sign-on banner on several public conferences and any course conferences observed, which indicates that:

This conference may be evaluated for preservation by the Bentley Historical Library for its potential to document the intellectual, cultural, and social environment of the University of Michigan.

This banner alerts participants in advance of entering text to the possibility that their contributions may be saved and later made available to researchers. It seems to be an agreeable compromise to participants on many of the privacy and ownership issues, as well as being in line with the spirit of the "Conditions of Use Statement." Project archivists also used discussion on conferences to their own advantage. Two conferences, with strong participation by many experienced conference users and computing center staff, were specifically established to discuss university computing issues, including those associated with the development, use, and organization of conferencing. Project archivists initiated items on these conferences to generate ideas as to the technicalities of how to preserve conference materials in the long term. This was particularly enlightening with regard to opinions about and institutional support for tape cartridge storage and routine backup procedures for electronic communications.

As with any project, certain events occurred that were beyond the control of the archivists and that necessitated a modification of original plans. In this case, developments within the university's Information Technology Division-a switch from two host mainframes to one and policy changes designed to reduce technical support to magnetic tape users and encourage the use of cartridges instead-required the project archivists to become involved with accessioning issues earlier in the project than had been planned in the original grant proposal.

As a result, project archivists accessioned electronic versions of the current volumes of MREV: FORUM, a conference sponsored by the editorial board of the Michigan Review, a conservative student publication, and LGM:RAP a conference discussing gay and lesbian issues. Negotiations continue for accessioning Wing: Span, a women's conference, and several conferences sponsored by the Information Technology Division. Project archivists are also negotiating with Robert Parnes to accession the private conference he created with several colleagues as a forum for discussion of the development and testing of enhancements to the conferencing software; a conference in some ways analogous to the scientist's lab notebook. To facilitate the accessioning of these conferences, the project archivists devised two new transfer agreements, one for public conferences that reflects the Bentley's university records policy, the other for private conferences that reflects its procedures for accepting personal materials from donors. These transfer agreements also raise the issue, however, of how to determine ownership of conferences and the legal status of private conferences using university computing resources.

FINDINGS AND RECOMMENDATIONS

Appraisal

From a compiled union list of 259 individual conferences (some of which no longer exist in any format), project staff were able to gain access to and appraise sixty-four. The appraisal recommendations fell into four categories: 1) accession in electronic format, 2) accession in whole or in part in paper format, 3) do not accession at present, but continue to monitor conference for possible reappraisal if the content changes with a new volume or new participants, and 4) do not accession or monitor farther.

Conference Appraisal Recommendations		
Recommendation	No.	%
accession in electronic format	18	28
accession in paper format	5	8
continue to monitor	12	18
do not accession or monitor	30	46
	65	100

The total of twenty-three conferences (thirty-five per cent) recommended to be accessioned exceeded the hypothesized ten per cent figure established for electronic conferencing to be considered a medium with archival value. It must be noted, however, that the high percentage of conferences recommended for accessioning is in part an artifact of a somewhat biased sample. The public conferences identified through the listing maintained by the Computing Center did constitute a self-selected sample of presumably important or popular conferences. Had all 259 of the conferences identified in the union list been examined, the percentage recommended for accessioning would have been significantly lower, but still well above the ten per cent threshold.

The appraisal process showed that computer conferences as a genre do indeed have potential to document the academic environment, and that, at least in the case of the University of Michigan, several individual conferences contain information which is unique or which significantly supplements traditional sources of archival information in the areas of intellectual history, pedagogy, and academic life and culture. One can say unequivocally that conference material is more current, directly reactive, and topical than traditional collections, since the immediacy of the medium makes them very responsive to current events and cultural trends. Subjects that came up again and again, although discussed from many different perspectives by the different participant communities of the various conferences, include race and race relations (such as racism on campus or the Rodney King verdict), gender issues (especially feminism and homosexuality), national and campus politics (for example, the Supreme Court nomination of Clarence Thomas or the presidential race, and the introduction of an armed campus police force); reproduction, health, and nutrition; children; recycling, alternative energy sources, and environmental conservation; evangelical religion; and role-playing games. Individual discussion items on conferences, such as those relating to "political correctness" or campus diversity can be very extensive, wide-ranging, and thoughtful, and thus have very evident research value. This value is further enhanced, however, both by the dynamics of the conference as a whole and by comparison with the differing perspectives of other conferences discussing the same issues. The value is also enhanced by the digital format of the materials, which will afford researchers opportunities for electronic content analysis not previously available with traditional textual university materials.

Where there were qualitative similarities between conferences and existing Bentley holdings, they were most frequently found to be with manuscript collections contained in the Michigan Historical Collections-the library's manuscript division-rather than with records of the University Archives and Records Programme (for example, discussions on issues such as abortion, political affairs, or environmental activism). This appears to be because of their personal nature and concern with issues that are germane not only to the university environment but also reflect upon community, national, and international concerns.

Those conferences with clear provenantial relationships to university units (administrative divisions, academic departments) or student organizations will be integrated with existing record groups. Conferences sponsored or owned by organizations, university units, or individuals for which the library does not have an existing record group or collection will be accessioned as independent record groups.

As stated earlier, in comparing the electronic conferences with the existing collections, the project archivists looked for overlapping coverage. They found that there was some subject and a very small amount of span-date overlap, but that qualitatively the electronic and paper records were yielding a very different documentary perspective.

The archivists looked also for existing collection strengths that would be enhanced significantly by those qualitatively different conferences; they found that for areas such as public health, feminism, and academic freedom (especially Political Correctness) the conferences did indeed provide significant new documentation. It was also important to look at areas documented by conferences that pointed out gaps in the Bentley's collection efforts. Among the gaps that conferences might fill at Michigan are those relating to documenting an aging student population and the challenging of traditional sex roles; topics made evident in frequent and extensive discussions of sexual harassment, homosexuality and parenting?

A particularly important new area of documentation made evident in the computer conferences is that of the impact of the computer on academic (especially student) life and culture. On many college and university campuses computers have become ubiquitous and linked network systems have become increasingly important. Academia is in the vanguard of the use of electronic communication for more than just administrative and research purposes. Electronic communication provides many different forums that can be used for learning, invisible college networking, debating, protesting, gossiping, grumbling, and partying--each of which develops a certain "culture" with norms for etiquette, language, and punctuation.

In many colleges and universities, there has been a sea-change in the campus environment due to an influx of students who are computer literate and many of whom own computers. This change in the student population as well as in networking capabilities has brought the computer sub-culture that had existed since the sixties into the nineties as an integral part of everyday student life. Few if any college and university archives have traditional holdings in non-electronic format that document well the computer revolution of the 1960s onwards and, in particular, the microcomputer and networks revolution of the 1980s and 1990s as they exert an impact upon the campus environment and youth in general.

One more interesting finding is a set of topics that were seldom if ever discussed on the computer conferences examined. These include fraternities and sororities, most student clubs, and college athletics and sports in general. These are often amongst the most highly documented topics in college and university archives, suggesting that in this sense too, traditional and electronic conferencing materials are complementary.

Comparing the conferencing with non-electronic materials was not as simple as the project archivists had envisaged. There was a problem associated with the contemporaneity of conferences: overlapping paper records might perhaps be available at some point in the future, but had not yet been accessioned or appraised. Where there were related materials within existing collections, they were sometimes described under more generic (or institutionally preferred) subject headings suitable for describing the entire collection. In comparing Library of Congress subject headings assigned to each material-type, there were also problems with matching dated and revised subject headings.

What the appraisal figures clearly demonstrate is that conferences must be appraised as distinct series and not globally as a format. There is no one strategy that is appropriate for all conferences; appraisal of active conferences needs to be conducted by college and university archivists on an ongoing basis, since individual conferences come and go and their nature can be quite different from year to year as organizers and student participants change. In addition, conferences may periodically be closed and restarted as a new "volume" if their size has reached the technical maximum allowed by the software, or if the conferences are run on an annual academic cycle. This ongoing appraisal can be done either as part of an archival records management programme, as manuscripts field work, or possibly as both.

ACCESSIONING AND STORAGE

One of the recommendations of the project was that the electronic records of conferences having archival value be retained in their original, software-dependent electronic format, while being frozen by the electronic "archiving" process in such a way that it would be almost impossible to tamper with the original record. To do this involves relatively simple "archiving" and remounting procedures on the part of the archivists and the organizers, for which written guidelines already exist. This format retains most of the evidential value of active conferences by capturing the dynamics and different dimensions of conferences in a way that it would be difficult for a researcher to reconstruct using a printed version of the text they contain. It also permits digital manipulation, either by the archivist to strip out and/or replace personal identifiers where required or by the researcher. Although there are important preservation concerns (mostly the question of software obsolescence), this method is efficient with space (a printout might run to several thousands of pages) as well as being most theoretically sound (since the dynamic elements of a conference cannot be transferred successfully to another medium). To this end, the Bentley has established an ongoing mainframe computer account with sufficient disk space allocation to be able to conduct electronic "archiving" of accessioned conferences, and to remount them online for research access. By selecting this option, however, archivists still retain the other storage options of generating "flat" files of text or of printing out conferences if at some point it becomes infeasible or not technically possible to migrate the conferences to an upgraded system.

Areas requiring further work and study

Issues relating to intellectual and physical access to "archived conferences represent the most extensive and pressing areas requiring further work and research by the Bentley. However, apart from committing to ongoing appraisal and accessioning of electronic conferences, some other areas still require further attention.

The appraisal work conducted to date should be codified into appraisal guidelines that could be used by non-technical archivists working with electronic conferences at other institutions. Since much of this appraisal work is laborious, and not always easy to assess manually, there should be further investigation into developing an automated front-end application that might assist the appraisal archivist in discerning use patterns and changes in them.26 Such information would greatly improve the archivist's ability to make informed appraisal decisions and develop accurate scope and content statements.

Further research needs to be conducted into the legal issues associated with literary rights of conference participants and with the definition of ownership of a conference. This is necessary to determine what is and is not a university (and therefore a public) record, so that informed decisions can be made when developing transfer donor agreements and access policies. Further research is also required to establish more precisely the nature and needs of potential user communities. Much more effort will need to be expended in outreach activities in order to bring those research communities to the conference materials.

Research Implications for the Profession

Two major questions must be posed by archivists looking at computer conferencing and any other form of electronic communication: to what extent will these electronic media have a long-term existence as distinct genres, and what will their record status be. Perhaps computer conferencing is a genre that is a product of just such a transition into the era of computer-mediated communication and cannot be relied upon to be in place twenty years from now. However, it was one of the earliest communication formats and has stayed the course for twenty years. It has also expanded from its original task-oriented role into a broader sociological phenomenon. Even if the genre were transient or non-record, to what extent would that devalue its documentary potential in the college and university environment with its multiple documentary mission? In David Bearman's words, "some written forms of electronic communication, such as intra-office uses of electronic mail, are currently undergoing cultural definition, and could come to be perceived as private in the absence of institutionally defined etiquettes and records policies." Bearman feels that such policies are central to effective electronic records management.?' The experience of the Bentley project archivists is that this approach is somewhat simplistic when one tries to implement it in an electronic communications environment that is loosely regulated and often subject to competing concerns of intellectual and academic freedom, personal privacy, and public disclosure laws.

While it is potentially dangerous to attempt to extrapolate standard approaches from one limited experience with only one of a myriad of communications media, the project archivists think that there are elements in their experiences and findings that do translate to generic approaches or raise generic issues in the electronic communication environment. For example, discussions of FOIA implications, authentication, ownership, preservation, description, access, custody, and appraisal mechanisms should be similar for all electronic communication-although they may not all play out the same way in different institutional environments. There is only one way to test if the approaches and conclusions of the Bentley Computer Conferencing Project have validity and generalizability in the college and university setting and the wider archival arena. That is to replicate them in other settings: a large university with a different student population, an undergraduate college, a corporation or non-profit institution.

What is of most critical importance for the profession at this juncture, however, is to develop a profile of the archival nature of electronic communication and examine further what has been learned here that might be of use to archivists exploring other forms of electronic communication. Well suited to the academic environment, perhaps, and certainly to the documentary goal of the Bentley's computer conferencing project-to look at computer conferencing as a means of documenting academic life-are the insights of Hugh Taylor on communication media, especially when they are in a period of transition. Taylor draws attention to some of the more subliminal aspects of electronic media by discussing the underlying symbols and messages beneath the text, and the need for archivists to be able to read them:

Archivists reared in a largely textual environment have had a tendency to "read" all media of record literally, without realizing that all forms of communication are loaded with conventions and semiotic "signs" inherent in their respective technologies. Consequently, archivists and users alike have to employ more perceptive strategies of interpretation.'

This project has made an important first step toward employing such strategies.

REFERENCES

1. Helen W. Samuels, "North American Archival Identity," in Judith A. Koucky, ed., Proceedings of the Second European Conference on Archives (Ann Arbor, 1989). p. 85; Varsity Letters: Documenting Modern Colleges and Universities (Metuchen, 1992).

2. Helen Horowitz, Campus Life: Understanding Undergraduate Cultures From the End of the Eighteenth Century to the Present (New York, 1987). is one of the few scholarly surveys. John Straw, "From Classroom to Commons: Documenting the Total Student Experience in Higher Education,"Archival Issues 19, no. 1 (Spring 1994).

3. Other forms of electronic communication in use in college and university environments include bulletin boards, listservs, Usenet newsgroups, and electronic mail. In many respects, computer conferencing permits more complex interactions than other forms of electronic communication.

4. See, for example, Andrew T. Finn, "Process and Structure in Computer-

Mediated Group Communication," in Brent D. Rubin, ed., Information and Behavior, Vol. II (New Brunswick, 1988), pp. 167-93; S.R. Hiltz and Murray Turoff, "The Evolution of User Behavior in a Computerized Conferencing System," Communications of the ACM 24 (November 1981); Ronald E. Rice, "Computer Conferencing" in Brenda Dervin and Melvin J. Voight, eds., Progress in Communication Sciences, Vol. I1 (Norwood, 1980), pp. 216-40; Jane Siegel, Vitaly Dubrowsky, Sara Kiesler, and Timothy McGuire, "Group Processes in Computer-Mediated Communication," Organizational Behavior and Human Decision Processes 37 (1986). pp. 157-87; Howard Rosenbaum and Herbert Snyder, "An Investigation of Emerging Norms in Computer Mediated Communication: An Empirical Study of Computer Conferencing," Proceedings of the 54th ASIS Annual Meeting, Washington, D.C., 27-31 October 1991, Jose-Marie Griffiths, ed., Learned Information, Inc. (N.J., 1991); Robin Mason, "Moderating Educational Computer Conferencing," Distance Education Online Symposium (DEOS) News 1, no. 19 (199 1); Terje Rasmussen, Joergen Bang, and Knut Lundby, "When Academia Goes Online: A Social Experiment with Electronic Conferencing for the Nordic Media Research Community," DEOSNEWS 1, no. 24 (1991); and Mary Joan Tooey, "Computer Conferencing: A Campus Goes Online," Online (July 1989), pp. 54, 57-60.

5.   "Penn State Electronic Records Appraisal Programme, Final Report," November 1993.

6.   It is widely used, for example, at the University of California campuses, the University of Maryland, Syracuse University, Rensselaer Polytechnic Institute, Carleton College, and the University of British Columbia in Canada.

7.   Ellen M. Pearson, "Computer Conferencing for Enhanced Communication: Its Potential for Academic and Research Communities" in Ahmed H. Helal and Joachim W. Weiss, eds., International Library Cooperation: 10th Anniversary Essen Symposium, 19 October - 22 October, 1987 (Essen, 1988). pp. 328-37.

8.   Matthew Rapaport, Computer Mediated Communications: Bulletin Boards, Computer Conferencing, Electronic Mail and Information Retrieval (New York, 1991), pp. 1-1 1 and passim, describes the evolution of conferencing and the development of the principal software packages.

9.   See University of Michigan Center for Research on Learning and Teaching, Records, Box 3, Bentley Historical Library; and Robert Parnes, "Learning to Confer: The Interplay of Theory and Practice in Computer Conferencing," (Ph.D. Dissertation, University of Michigan,

198 1). According to Parnes, "The ideas which became CONFER had their origins and early development in a seminar led by Merrill Flood on problems in academic governance .... The seminar took place in the School of Education at The University of Michigan in the fall of 1974. Its goal was to examine ways of improving both the communications and decis~on making processes as they related to academic governance. An underlying assumption was that if any important impact were to be hoped for, something significantly different from the status quo, i.e., a new technology, would need to be developed. Individual interest in computers led to a focus on how they might be used in this context. Merrill Flood set about examining the decision making process and I the communications process. It was the intention of both of us to integrate the results of our work into a system making other features and facilities that allow participants to choose their own styles and levels of involvement." (Parnes, "Learning to Confer," pp. 40-41).

10.   CONFER and CONFER 11 are registered trademarks of Advertel Communications Systems. For discussions of conferencing systems and comparisons of conferencing software see Elaine B. Kerr and Roxanne Starr Hiltz, Computer-Mediated Communication Systems: Status and Evaluation (New York, 1982); and Rapaport, Computer Mediated Communications.

11.   One of the factors that assisted in the promotion and spread of CONFER was that the Michigan Terminal System (MTS), an operating system developed in 1967 at the University of Michigan to run on IBM and compatible mainframe computers, and upon which CONFER resides (although it can function independently of MTS also), had been sold internationally. Other purchasers have included the University of Illinois, Rensselaer Polytechnic Institute, the National Laboratory for Scientific Computing in Rio de Janeiro, the University of British Columbia in Canada, and Durham University and the University of Newcastle Upon Tyne in England. CONFER I1 has been ported to the DEC VMS and Unix operating system.

12.   Electronic mail communication from Robert Parnes to Carol Hughes and Anne Gilliland-Swetland, 18 October 199 1.

13.   Parnes, "Learning to Confer," p. 1 18.

14.   Important recent publications for academic archivists contemplating the archival administration of electronic records include Charles Dollar, Archival Theory and Information Technologies: The Impact of Infornzation Technologies on Archival Principles and Method (Macerata, 1992); David Bearman, Electronic Records Guidelines: A Manual for

Policy Development and Implementation (Pittsburgh, 1990); United Nations Advisory Committee for the Coordination of Information Systems, Electronic Records Guidelines. A Manual for Policy Development and Implementation (New York and Geneva, 1989); Katharine Gavrel, Conceptual Problems Posed hv Electronic Records: A RAMP Study (Paris, April 1990); Margaret Hedstrom, "Privacy, Computers, and Research Access to Confidential Information," Midwestern Archivist 6, no. 1 (1981). pp. 5-18; and "Understanding Electronic Incunabula: A Framework for Research on Electronic Records," American Archivist 54 (Summer 1991): pp. 334-54. Two works, Margaret Hedstrom, Archives and Manu.script.s: Machine-Readable Records (Chicago, 1984) and Harold Naugler, The Archival Appraisal of Machine-Readable Records: A RAMP Study with Guidelines (Paris, November 1984). are very dated, but still provide information useful to archivists on systems background and offer procedural advice not found in other texts.

15. The University and the Computing Center have been concerned to establish and maintain (and the user community has demanded) an open and free computing environment with minimal monitoring of individual's computing activity. It is felt that requiring reporting and maintainance of metadata on all conferences might inhibit the free tlow of ideas in an academic setting. These concerns apply particularly to "private" conferences.

16. Parnes, "Learning to Confer," pp. 41-42. Parnes has stated that "CONFER was designed-both consciously and unconsciously-to incorporate the liberal democratic values of equality of opportunity and civil liberties," and "As a consequence of the values of equality and freedom embodied in CONFER, there can result a tendency towards freewheeling, extended discussion. CONFER makes possible unfocused, open-ended discussion, and is somewhat biased towards allowing people to write too much, rather than too little."

17. This is a concern that has also been voiced by archivists, albeit probably not thinking about an example as specific as computer conferencing. Katharine Gavrel, for example, has written that "the difficulty from an archival perspective on the early involvement of the archivist in the development and design stage of any electronic system, be it an office system or a database, is the influence the archivist will have over the records being created and identified for long term preservation. Archives have traditionally been the passive receivers of documents, making appraisal decisions at the end of the active life of the document. If archives become active participants in the development process, what

influence will that have over the information that is being created and designated as archivally valuable?' (Gavrel, Conceptual Problems, pp. 27-28).

18.   Parnes, "Learning to Confer," pp. 66-67.

19.   The university had operated two mainframe systems, designated "UB." which hosted most student accounts, and "UM which hosted most faculty and staff accounts. Many people had accounts on both systems. When UB was phased out in January 1992 conferences on that system could elect to move intact to UM or close their current UB volume and restart on UM. Since completion of the project, the university has begun to move away from its mainframe to a distrubuted computing environment, which spurred Parries's develoment of a Unix based version of CONFER known as CONFER-U. This poses additional problems for archivists, as the text, participation, and other files of individual conferences will no longer be resident on a single piece of hardware but possibly spread over many machines.

20.   Organizers wishing to "archive" a computer conference originally did so on magnetic reel tape. At present, cartridge tapes are the preferred method of data storage. Cartridges offer a potentially more stable storage medium than reel tape, and since the late 1980s have come down considerably in price, albeit that they are still not a long-term preservation medium in the archival sense. See Thomas E. Weir, Jr., 3480 Class Tape Cartridge Drives and Archival Data Storage: Technology Assessment Report, National Archives Technical Information Paper No. 4 (Washington, D.C., June 1988).

21.   Accessioning in paper format was recommended for three reasons: 1) the electronic version no longer existed, 2) the conference was very small, or 3) only a few selected items from a conference were judged to have archival value, in which case it was deemed most efficient to print those items and add them to existing record groups as new series.

22.   It should also be noted that several of the course conferences observed were judged to have archival value, but inability to comply with FERPA regulations resulted in a recommendation not to accession. See Anne Gilliland-Swetland and Gregory Kinney, "Uses of Electronic Communication to Document an Academic Community, Appendix 11," Final Report to the NHPRC, December 1992.

23.   In the "archiving" process the text of the conference is "frozen" insuring the integrity of the original as it was received by the archives. This still permits archivists and researchers to use CONFER 11's various index, search, and query features and to download or print portions of text.

Duplicate copies of tapes would be generated for research use. The conference file is "renderable" and "redactable" to use Bearman's terms. David Bearman, "Record-Keeping Systems," Archivaria 36 (Autumn 1993). p. 32.

24.	Anne J. Gilliland-Swetland and Carol Hughes, "Enhancing Archival Description for Public Computer Conferences of Historical Value: An Exploratory Study," American Archivist 55 (Spring l992), pp. 3 16-30,

25.	Bearman, Electronic Records Guidelines, p. 17.

26.	Hugh Taylor, "'My Very Act and Deed': Some Reflections on the Role of Textual Records in the Conduct of Affairs," American Archivist 5 1 (Fall 1988). p. 457.

Chapter 4

FEASIBILITY STUDY ON A PORTABLE FIELD PEST CLASSIFICATION SYSTEM DESIGN BASED ON DSP AND 3G WIRELESS COMMUNICATION TECHNOLOGY

Ruizhen Han [1,2], Yong He [1], and Fei Liu [1]

[1]College of Biosystems Engineering and Food Science, Zhejiang University, Hangzhou 310058, China

[2]College of Electronic Information, Zhejiang University of Media and Communications, Hangzhou 310018, China

ABSTRACT

This paper presents a feasibility study on a real-time in field pest classification system design based on Blackfin DSP and 3G wireless communication technology. This prototype system is composed of remote on-line classification platform (ROCP), which uses a digital signal processor (DSP) as a core CPU, and a host control platform (HCP). The ROCP is in charge of acquiring the pest image, extracting image features and detecting the class of pest using an Artificial Neural Network (ANN) classifier. It sends the image data, which is encoded using JPEG 2000 in DSP, to the HCP through the 3G network at the same time for further identification. The image transmission and communication are accomplished using 3G technology. Our system transmits the data via a commercial base station. The system can work properly based on the effective coverage of base stations, no matter the distance from the ROCP to the HCP. In the HCP, the image data is decoded and the pest image displayed in real-time for further identification. Authentication and performance tests of the prototype system were conducted. The authentication test showed that the image data were transmitted correctly. Based on the performance test results on six classes of pests, the average accuracy is 82%. Considering the different live pests' pose and different field lighting conditions, the result is satisfactory. The proposed technique is well suited for implementation in field pest classification on-line for precision agriculture.

INTRODUCTION

Pest control has always been considered the most difficult challenge to overcome in agriculture. Traditionally, pest management has been accomplished by means of a regular spray program which is based on a schedule rather than on the presence or likelihood of presence of insects in the field. More recently, growers have incorporated weather-based models to predict pest presence and apply control methods based on these models [1]. The most accurate method to control pests, and a method which is gaining interest in the wake of the need to minimize environmental impacts, is integrated pest management (IPM). The four main steps of IPM are detection, identification, application of the correct management and registration of the management [2]. The primary challenge with those steps is the identification. Classification of insect species can be extremely time consuming and requires technical expertise, so an automated insect identification method is needed. Due to the rapid development of digital image technology, there is a growing tendency in the field of agricultural research towards using machine vision technology to help the research and solve problems. In recent years, the use of artificial neural networks (ANN) has spread to many branches of science. Image analysis and ANN provide a realistic opportunity for the automation of routine species identification [3]. Do et al. [4] utilized various artificial neural networks to identify spider species using only digital images of female genitalia and achieved an average species accuracy of 81%. Artificial neural networks based on morphometric characters have been already applied in insect identification. Vanhara et al. [5] tested the methodology of ANN identification in the family Tachinidae on the basis of five model species of two genera, using 16 morphometric characters. Fedor et al. [6,7] identified Thysanoptera species using artificial neural networks with the morphometric characters. Russell et al. [8] developed an on-line automated identification system called SPIDA. The SPIDA system is trained to identify the 121 species of the Australasian spider family Trochanteriidae based on an artificial neural network model. SPIDA is currently available on the Internet, and users can submit their own images of spiders for classification, although some expertise and equipment is required to obtain optimal images. Murarikova et al. [9] confirmed the power of ANN by two independent non-numerical methods (molecular analysis, comparative morphology).

Most of the existing systems are semi-automated and all these systems have been trained on images taken from dead specimens. In a laboratory, dead specimens can be carefully positioned and photographed under consistent and ideal lighting conditions. In the field, however, live specimens may not adopt the ideal pose required, they may move when the image is being captured, and the lighting conditions outside the lab may be poor and may change unpredictably as a series of images are taken. This tends to make the classification task much more difficult.

Mayo and Watson [10] described different classifiers and datasets to identify live moths automatically and indicated that the best classifier is Support Vector Machine which achieved approximately 85% accuracy without manual preprocessing of the images. However, in those systems, the process of the training and testing was done in the laboratory and they can't classify insects in real-time on-site. In order to detect the insects earlier, we aim to develop an on-line automated live insect identification system, which is portable and can provide the classification results in the field. The main research objectives of this paper were:

To design a hardware platform to implement image capturing, image processing, pest classifying with an Artificial Neural Network (ANN) classifier and image encoding.

- To process the images and to conduct pest classification in DSP.
- To design a wireless communication protocol and to transmit the images with a 3G network.
- To display and store the pest images for expert precise classification and to design a host control platform for completing image decoding.
- To test the designed system in the field.

This paper is organized as follows: Section 2 discusses the principles and algorithm flow of artificial neural network (ANN). Section 3 presents the hardware and software design of this system. Section 4 is devoted to the test of the proposed system. Field test results are provided in Section 5. Finally, conclusions are drawn in Section 6.

ARTIFICIAL NEURAL NETWORK

Artificial neural networks (ANNs) provide a way to emulate biological neurons to solve complex problems in the same manner to the human brain. For many years, especially since the middle of the last century, interest in studying the mechanism and structure of the brain has been increasing. In 1986, the Parallel Distributed Processing (PDP) research group published a series of algorithms and results and presented an ANN training algorithm named Back Propagation (BP) [11,12]. This BP training algorithm implemented with the general delta rule gave a strong impulse to subsequent research and resulted in the largest body of research and applications in ANNs although many other ANN architectures and training algorithms have been developed and applied simultaneously.

The massively parallel architecture of the ANN consists of multiple layers of simple computing elements with many interconnections between the layers. The computing elements are functionally analogous to neurons. They receive signals and in turn transmit a signal which is a function of the inputs. The function by which the inputs are evaluated may be a simple logic gate but more generally involves summation of weighted input signals. A transfer function is then applied to the weighted inputs to determine the output of the neuron. In this paper, we used a three-layer BP-ANN. Figure 1shows the feedforward network between input X and output Y. In this paper, the BP-ANN was trained in advance via large numbers of experimental data. This training process was accomplished using Matlab language on a PC. After the BP-ANN was trained, the weights and thresholds were programmed in DSP for the BP-ANN model.

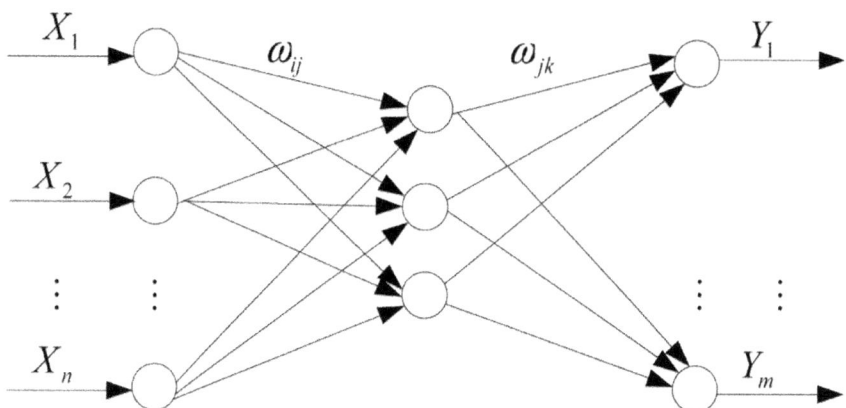

Figure 1: ANN network structure.

PROTOTYPE SYSTEM DESIGN AND IMPLEMENTATION

Hardware Design

The prototype system architecture adopted in this work is shown in Figure 2. This system includes a remote on-line classification platform (ROCP) and a host control platform (HCP). The ROCP mainly consists of a DSP, a 3G network module, an image sensor module, a LCD module and a power module. The HCP is composed of a PC and a modem for accessing the internet. The HCP can receive the image data send by the ROCP, decode them and display the image.

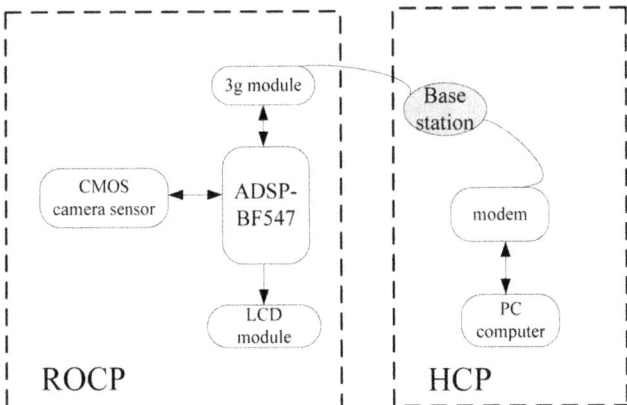

Figure 2: Architecture of the designed testing system.

With the image sensor, we can get the pest images to the DSP in the ROCP platform. The DSP has two important things to do: on the one hand, it will preprocess the images, compute the features of the images, and give the initial classification results obtained from the BP-ANN classifier. On the other hand, it will encode the image data using JPEG 2000 and send them to the HCP through the 3G module. After receiving those image data, the HCP will then decode these data and display the pest images.

This prototype system utilized an ADSP-BF547 processor as a kernel CPU in ROCP platform. The ADSP-BF547 processor is a member of the Blackfin family of products, incorporating the Analog Devices, Inc./Intel Micro Signal Architecture (MSA). The processor core clock is up to 600 MHz. It's Dynamic Power Management provides the control functions to dynamically alter the processor core supply voltage to further reduce power consumption. Control of clocking to each of the peripherals also reduces power consumption. This is very suitable for portable appliances. The ADSP-BF547 processor peripherals include three SPI ports, eleven general-purpose timers with PWM capability, a real-time clock, a watchdog timer, a parallel peripheral interface, which is connected with the image sensor, an enhanced parallel peripheral interface which is connected with LCD module, and four UART ports, one of them is used to connect with the 3G module (module no: SIM5218A) for data transmission. The CMOS camera module (module no: OV9650) is used for pest image acquisition, the OV9650 is a color image sensor and has 1.3-Mpixel which is suitable considering the hardware resource and image resolution. Figure 3 shows photographs of the designed system.

CMOS camera module   DSP module   3G network module

 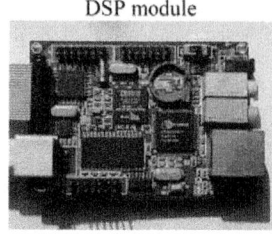 

Figure 3: Photographs of the designed system including DSP module, CMOS camera module and 3G module.

Software Development

According to the hardware architecture of the designed portable system, the tasks of the whole system are the pest classification on DSP, image data compression coding, wireless data transmission, image decompression and image display on a PC. Therefore, software development of the system includes two parts—DSP software design and PC software design. The DSP programs are designed in three steps. Firstly, the data acquisition program acquires the image sensor response data.

Secondly, DSP processes the image data, extracts the features and provides the classification results. Finally, it encodes the image data using JPEG 2000, packages them into different frames and sends them to a PC with the 3G module. The specific program flow diagram is shown in Figure 4. The image preprocessing is composed of image transforming, threshold processing, binarization and denoising. After finishing the image preprocessing, we extracted the image's morphological characteristics including eccentricity ratio, sphericity and two Hu invariant moments for classification.

In addition, we designed a wireless communication protocol and used the universal asynchronous receiver/transmitter (UART) interfaces of the DSP to carry out the serial data transmission between the 3G module and the DSP. The data frame format is composed of a frame head ($0\times1B$, $0\times7E$), sequence number (two bytes), valid data bytes, and frame end ($0\times FF$), as shown in Figure 5. Each frame has 512 valid data bytes. Communication baud rate is set at 57,600 baud. The flow diagram of the communication program of DSP is shown in Figure 6.

After establishing the TCP/IP connection, we started to send the data and enable the timer which is used for avoid the system halting because of no return from the 3G module at the same time. If the timer expired and returned nothing, we resend the same data again. If the returned information is errorroneous, we reset the 3G module and establish the TCP/IP connection again. If we receive the right reply, we send next frame data until all data are sent.

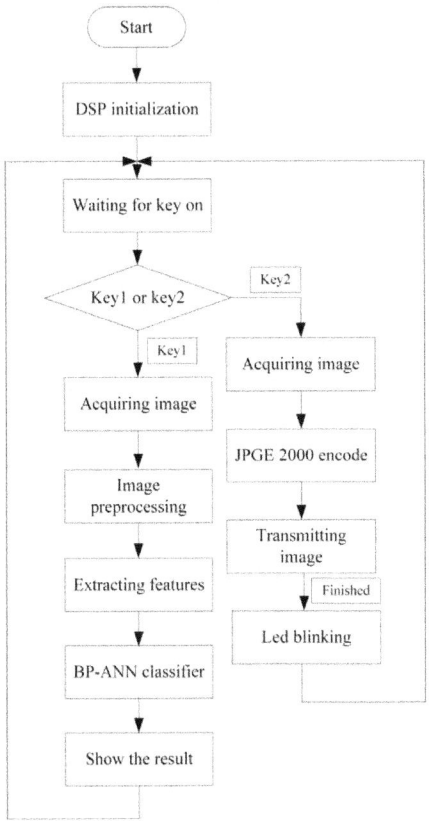

Figure 4: Flow diagram of the DSP program.

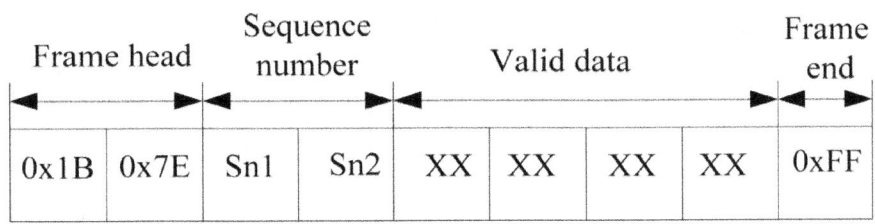

Figure 5: Wireless communication protocol.

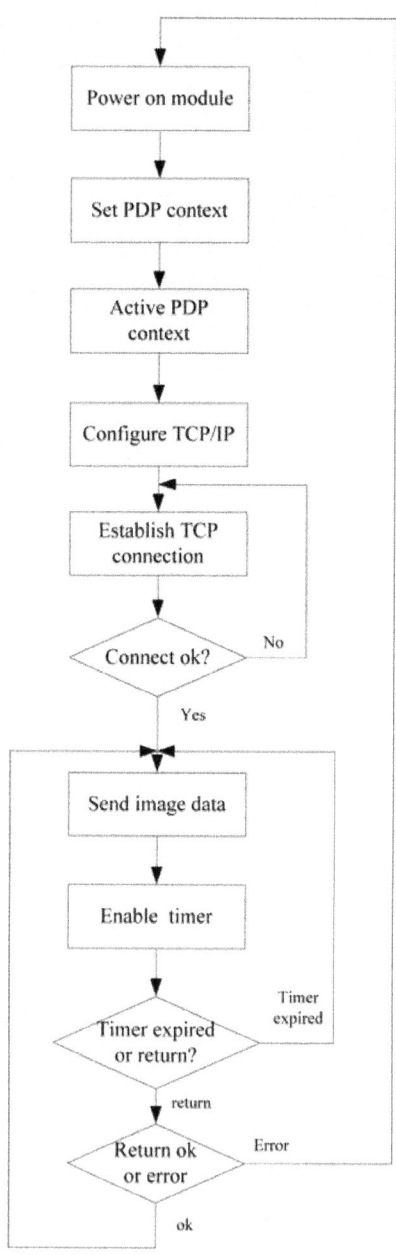

Figure 6: Wireless communication process.

The PC software of the HCP programmed in Visual C++ language decodes the image data, displays the images and stores the images.

FEASIBILITY STUDY OF THE DESIGNED PORTABLE SYS-TEM

The feasibility study of the designed system was composed of three sections: DSP image acquisition and image processing tested the effect of image processing algorithm and extracted the morphology and color features. By training the BP-ANN, we obtained the weights and thresholds of the BP-ANN model. The data transmission authentication test validated the reliability of 3G network transmission.

DSP Image Acquisition and Image Processing Test

The dataset used in this study is a library of live pest images created by the first author over a period of nearly a year. A pest trap was set up in the Fuyang Plant Protection Station (Zhejiang Province, China) and cleared every morning. Captured live pests were photographed and then released. Cnaphalocrocis medinalis Guenee is taken as an example and the image is shown in Figure 7.

Figure 7: Image of Cnaphalocrocis medinalis Guenee.

Considering that the trapped field pest's morphological characteristics and color have relatively large differences, we extracted the morphology features and color features for classification. Geometrical features which describe the geometric properties of the target area are unrelated to the color value of the region. Therefore, the image is binarized before extracting it's geometrical features. The Figure 8depicts the automatic processing and feature extraction pipeline, using Figure 7 as an example input. The first step in feature extraction was to transform from the RGB color space to the HSV color space.Figure 8(a) depicts the results of the H-component when applied to the image of Cnaphalocrocis medinalis Guenee in Figure 7. The static threshold was obtained according to the statistics in the H-component, and was used for the input image and produced a threshold image as shown in Figure 8(b). Then the threshold image is binarized as shown in Figure 8(c). Finally, in order to reduce the noise, we adopted the method of searching the maximum linked area. We used the recursion method to find all connected region in which the value is "1", and compared their size. The largest of them is the target object, the other is the noise. The result is shown in Figure 8(d). Now, a number of morphology features were calculated. The color features were described by color moments [13]. All features consisted of nine color moments, eccentricity ratio, sphericity and two Hu invariant moments which are invariant to image scaling, rotation and translation [14]. The color moments are defined by the following equations:

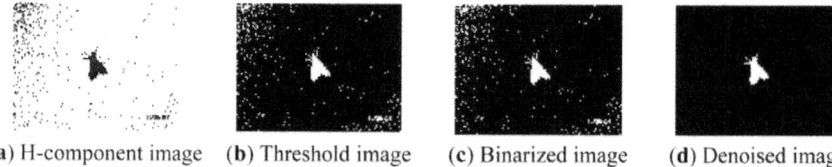

(a) H-component image (b) Threshold image (c) Binarized image (d) Denoised image

Figure 8: The image processing pipeline.

$$E_i = \frac{1}{N} \sum_{j=1}^{N} P_{ij} \tag{1}$$

$$\sigma_i = \left(\frac{1}{N} \sum_{j=1}^{N} (p_{ij} - E_i)^2 \right)^{\frac{1}{2}} \tag{2}$$

$$s_i = \left(\frac{1}{N} \sum_{j=1}^{N} (p_{ij} - E_i)^3 \right)^{\frac{1}{3}} \tag{3}$$

where P_{ij} is the value of the ith color channel at the jth image pixel, $i \in \{1, 2, 3\}$, N is the number of image pixel.

The eccentricity ratio and the sphericity are defined by Equation (4) and Equation (5) respectively [15]. Two Hu invariant moments are defined by Equations (6) and (7) [14]:

$$EC = p/q \tag{4}$$

where p and q are half the length of principal axis of momental ellipse:

$$SP = r_i / r_c \tag{5}$$

where r_i, r_c are the radius of the inscribed circle and the circumscribed circle of the target object respectively:

$$\varphi_1 = \eta_{20} + \eta_{02} \tag{6}$$

$$\varphi_2 = (\eta_{20} - \eta_{02})^2 + 4\eta_{11}^2 \tag{7}$$

where η_{pq} is the normalized central moments.

BP-ANN Model Training Process

The architecture of our BP-ANN was established according to the number of input neurons and the number of classifications. The initial ANN consisted of a layer of input neurons, a hidden-layer and a layer of output neurons, fully interconnected with the hidden-layer by random initial weights. Each input layer neuron corresponded to a feature. The number of nodes in the hidden-layer needs to be considered. As a preliminary selection, the optimum number of nodes in the hidden layer was determined by Equation (8) [16]:

$$n_1 = \sqrt{n+m} + a \qquad (8)$$

where n_1 is the number of nodes in the hidden layer, n is the number of input nodes, m is the number of output nodes, and a is an experiential integer from 1 to 10. By comparing the classification results of different models, we choose the model which has 13 input nodes, 10 nodes in the hidden-layer and 6 nodes in the output layer in the final model.

We selected six common field pests (Cnaphalocrocis medinalis Guenee, Chilo suppressalis, Sesamia inferens, Naranga aenescens Moore, Anomala cupripes Hope, Prodenia litura) for training the BP-ANN model. After acquiring and processing the images according to the procedure above from all samples, the set of images was divided into a training set and a test set. The composition of the training set is shown in Table 1. In order to remove the effects resulting from the difference of all features' dimension, all features are normalized using the Equation (9) [17]:

Table 1: Composition of the training sets used to train BP-ANN

Species	Number
Cnaphalocrocis medinalis Guenee	70
Chilo suppressalis	69
Sesamia inferens	72
Naranga aenescens Moore	70
Anomala cupripes Hope	75
Prodenia litura	76

$$x_{ij}' = \frac{x_{ij} - \min_j}{\max_j - \min_j} \qquad (i = 1,2,\cdots,n;\ j = 1,2,\cdots,d)$$

$$(9)$$

where n is the number of pests and d is the number of features; x_{ij} and x_{ij}' are the non-transformed data and transformed data of the jth feature of the ith pest.

max_j and min_j are the maximum and minimum of the jth feature in the all pests. After the training was finished, the test was done according to the trained BP-ANN model. The composition of the testing set and the testing result are shown in Table 2.

Table 2: Testing results for BP-ANN

Species	Number	Accuracy (%)
Cnaphalocrocis medinalis Guenee	23	83
Chilo suppressalis	15	80
Sesamia inferens	18	81
Naranga aenescens Moore	21	82
Anomala cupripes Hope	25	88
Prodenia litura	30	85
Overall	132	83

Data Transmission Authentication Test

This test was conducted to verify the accuracy of the data sent and received during image data acquisition. The whole prototype testing system was implemented and placed at Zhejiang University Digital Agriculture and Agriculture Information Technology Research Center for conducting the test via China Unicom's WCDMA network. Using the previously described test system and the transmission data format, the image data, after being processed by ADSP-BF547 which included in encoding, packaging, were sent to the HCP using the 3G wireless transmission module. Comparisons were performed to check transmission time and image data correctness. The received image data were stored on the PC hard disk. The actually image data were read from the SDRAM in ROCP. These data were then compared with the image data received by the HCP subsystem using the ultraedit software. The comparison shows that the received data are correct.

RESULTS

After obtaining the weights and thresholds of BP-ANN, we programmed them in DSP for identification in the field. The CMOS image sensor OV9650 is used for image acquisition. The test was done in Fuyang plant protection station (Zhejiang, China). The number of test samples and the test results are shown in Table 3.

Table 3: Testing results for BP-ANN in the field

Species	Number	Accuracy (%)
Cnaphalocrocis medinalis Guenee	20	82
Chilo suppressalis	18	79
Sesamia inferens	20	80
Naranga aenescens Moore	21	82
Anomala cupripes Hope	25	86
Prodenia litura	23	82
Overall	127	82

The performance of the trained BP-ANN in the testing runs demonstrated that the designed system was capable of identifying the common six pests, which were trapped at the Fuyang Plant Protection Station, with an overall average accuracy level of 82%. This level of accuracy was satisfactory given the complex field conditions and the limited amount of information on which the identification system was based. It should be emphasized that the set of analysed pests is rather a model example to demonstrate the potential of artificial intelligence in this area. In later work, we will study the identification of those pests which are not easily distinguishable by traditional taxonomic keys.

Figure 9 shows the GUI of the HCP; the photo, corresponding to the original image shown in Figure 7, is displayed by decompression. It is clear enough for experts to identify it.

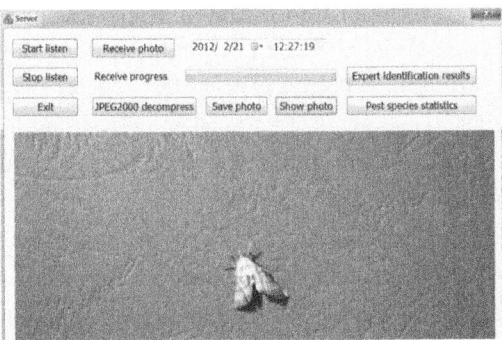

Figure 9: The GUI of HCP.

CONCLUSIONS

Due to the need for pest identification in the field for precision agriculture, this paper studied the feasibility of on-line pest classification using machine vision technology. A DSP was used for this due to its powerful data processing functions. Considering the complex situation related to the field and the resource limitations of DSP, the classification achieved is satisfactory. Image data, which was encoded with JPEG 2000, was transmitted through the WCDMA network to an HCP for further identification. The test results show that the DSP can provide an initial result and the pest image in HCP is very clear and sufficient for further identification. The design of a reliable automatic pest classification system in HCP will be the focus our subsequent research efforts.

ACKNOWLEDGMENTS

This study was supported by Important Zhejiang Provincial Science & Technology Specific Projects (2009C12002), Zhejiang Provincial Natural Science Foundation of China (Project No: Z3090295) and National Agricultural Science and Technology Achievements Transformation Fund Programs (2009GB23600517).

REFERENCES

1.   Jiang, J.-A.; Tseng, C.-L.; Lu, F.-M.; Yang, E.-C.; Wu, Z.-S.; Chen, C.-P.; Lin, S.-H.; Lin, K.-C.; Liao, C.-S. A GSM-based remote wireless automatic monitoring system for field information: A case study for ecological monitoring of the oriental fruit fly, Bactrocera dorsalis (Hendel).Comput. Electron. Agric 2008, 62, 243–259.

2.   Solis-Sanchez, L.O.; Castañeda-Miranda, R.; García-Escalante, J.J.; Torres-Pacheco, I.; Guevara-González, R.G.; Castañeda-Miranda, C.L.; Alaniz-Lumbreras, P.D. Scale invariant feature approach for insect monitoring. Comput. Electron. Agric 2001, 75, 92–99.

3.   Weeks, P.J.D.; Gaston, K.J. Image analysis, neural networks, and the taxonomic impediment to biodiversity studies. Biodivers. Conserv 1997, 6, 263–274.

4.   Do, M.T.; Harp, J.M.; Norris, K.C. A test of a pattern recognition system for identification of spiders. Bull. Entomol. Res 1999, 89, 217–224.

5.   Vanhara, J.; Murarikova, N.; Malenovsky, I.; Havel, J. Artificial Neural Networks for fly identification: A case study from the genera Tachina and Ectophasia (Diptera, Tachinidae). Biol. Bratisl 2007, 62, 462–469.

6.   Fedor, P.; Malenovsky, I.; Vanhara, J.; Sierka, W.; Havel, J. Thrips (Thysanoptera) identification using artificial neural networks. Bull. Entomol. Res 2008, 98, 437–447.

7.   Fedor, P.; Vanhara, J.; Havel, J.; Malenovsky, I.; Spellerberg, I. Artificial

intelligence in pest insect monitoring. Syst. Entomol 2009, 34, 398–400.

8.   Russell, K.N.; Do, M.T.; Platnick, N.I. Introducing SPIDA-Web: An Automated Identification System for Biological Species. Proceedings of Taxonomic Database Working Group Annual Meeting, St Petersburg, Russia, 11–18 September 2005.

9.   Murarikova, N.; Vanhara, J.; Tothova, A.; Havel, J. Polyphasic approach applying artificial neural networks, molecular analysis and postabdomen morphology to West Palaearctic Tachina spp. (Diptera, Tachinidae). Bull. Entomol. Res 2011, 101, 165–175.

10. Mayo, M.; Watson, A.T. Automatic species identification of live moths. Knowl. Based Syst 2007,20, 195–202.

11. Huang, Y.; Lan, Y.; Thomson, S.J.; Fang, A.; Hoffmann, W.C.; Lacey, R.E. Development of soft computing and applications in agricultural and biological engineering. Comput. Electron. Agric2010, 71, 107–127.

12. Wang, Y. Principle and Method of Artificial Intelligence, 1st ed; Xiꞌan Jiao Tong University Press: Xi'an, China, 1998; pp. 412–441.

13. Stricker, M.; Orengo, M. Similarity of color images. Proc. SPIE 1995, 2420, 381–392.

14. Wang, X.-F.; Huang, D.-S.; Du, J.-X.; Xu, H.; Heutte, L. Classification of plant leaf images with complicated background. Appl. Math. Comput 2008, 205, 916–926.

15. Zhang, H.; Mao, H.; Qiu, D. Feature extraction for the stored-grain insect detection system based on image recognition technology. Trans. CSAE 2009, 25, 126–130, (in Chinese with English abstract)..

16. Zhang, L. Models and Applications of Artificial Neural Networks; Fudan University Press: Shanghai, China, 1993; p. 46.

17. Darafsheh, M.R.; Moghaddamfar, A.R.; Zokayi, A.R. A recognition of simple groups psl(3, q) by their element orders. Acta Math. Sci 2004, 21B, 45–51.

Chapter 5

LOAD-ADAPTIVE PRACTICAL MULTI-CHANNEL COMMUNICATIONS IN WIRELESS SENSOR NETWORKS

Md. Shariful Islam, Muhammad Mahbub Alam, Choong Seon Hong and Sungwon Lee

Department of Computer Engineering, Kyung Hee University, 1 Seocheon, Giheung, Yongin, Gyeonggi 449-701, Korea

ABSTRACT

In recent years, a significant number of sensor node prototypes have been designed that provide communications in multiple channels. This multi-channel feature can be effectively exploited to increase the overall capacity and performance of wireless sensor networks (WSNs). In this paper, we present a multi-channel communications system for WSNs that is referred to as load-adaptive practical multi-channel communications (LPMC). LPMC estimates the active load of a channel at the sink since it has a more comprehensive view of the network behavior, and dynamically adds or removes channels based on the estimated load. LPMC updates the routing path to balance the loads of the channels. The nodes in a path use the same channel; therefore, they do not need to switch channels to receive or forward packets. LPMC has been evaluated through extensive simulations, and the results demonstrate that it can effectively increase the delivery ratio, network throughput, and channel utilization, and that it can decrease the end-to-end delay and energy consumption.

INTRODUCTION

A wireless sensor network consists of battery-powered sensing devices that transmit their observations to the base station. The sensing nodes have a limited transmission range, so nodes away from the base station deliver their data through intermediate nodes. The data generation rates of the sensing nodes depend on the applications. An elastic application might use varying data rates. For example, a monitoring application generates data at a very low rate in the absence of an event, whereas a particular feature might lead to a huge traffic burst [1]. Because of the limited capacity of nodes, the generated data often exceed the network capacity, leading to congestion and contention loss. A congestion control mechanism alleviates the congestion by restricting the nodes from generating data that the network cannot deliver. This ensures optimum usage of the resources and decreases congestion losses. However, if the application requires higher data rates, a congestion control (or rate control) mechanism cannot meet the demand. Therefore, some works have advocated for increased network resources to avoid congestion and to deliver the required data [2,3].

On the contrary, recent sensor motes, such as MicaZ [4] and Telos [5], are capable of using a number of channels [6]. A single adapter can use different channels at different times. If nearby nodes use orthogonal channels, multiple nodes can transmit simultaneously, thereby increasing the network capacity. Multi-channel communications can provide the required data delivery without adding extra resources. The use of a single channel is, therefore, not only an under-utilization of the limited resources of WSNs, but it might also hinder the fidelity of the application.

To improve the network capacity, many multi-channel medium access control (MAC) protocols have been proposed. These protocols generally assign (as part of the network setup) orthogonal channels to the nodes (either to the senders or the receivers) in a two-hop neighborhood [7–9]. The data transmissions among neighbors, therefore, require channel switching and a sophisticated MAC scheme to find a rendezvous time for the sender and receiver. As a result, such protocols require fine-grained time synchronization among the nodes.

To minimize the channel switching and to use multiple channels when necessary, a recent paper proposes a dynamic channel allocation policy based on control theory (hereafter referred to as DM-MAC) [10]. Because the nodes in DM-MAC change channels in a distributed manner for multihop communications, the nodes still need to switch channels. In order to completely avoid channel switching, a static channel allocation policy is proposed in TMCP [11]. TMCP divides the network into a number of sink-rooted sub-trees, where each sub-tree uses an orthogonal channel. However, the sub-tree creation requires a costly initialization phase. However, the sub-tree creation requires a costly initialization phase.

In this paper, we design a multi-channel communications system for WSNs. LPMC dynamically adds or removes channels based on the active network load, and uses multiple channels whenever (when the network load is higher than the capacity) and wherever (the part of the network with a high load) it is necessary. While LPMC has a similar flavor in terms of channel switching to DM-MAC [10], and in terms of channel allocation to TMCP [11], it differs considerably in the following ways:

Unlike TMCP, LPMC does not need any initialization, such that the overhead is reduced. LPMC assigns channels dynamically instead of the static allocation of TMCP. TMCP divides the network into sub-trees by considering the equal data rate of the nodes. Due to the dynamic channel allocation, LPMC is transparent to data rates.

Unlike DM-MAC, LPMC adds or removes channels based on the overall network load. Furthermore, the sink controls the channel changing instead of the sensor nodes, since it has a more comprehensive view of the overall network traffic.

The main contributions of this paper can be summarized as follows: (i) We propose a multi-channel communication systems for WSNs that keeps the protocol functionalities out of the sensor nodes as much as possible. (ii) LPMC dynamically identifies the network load and adds channel(s) to the mostly heavily loaded part. No initialization steps are required for LPMC, and the overhead for channel assignment is minimal. Nodes do not need to switch channels to receive or forward packets. (iii) LPMC dynamically adds paths with non-interfering channels to a set of nodes to meet the traffic demands. (iv) The performance of LPMC is evaluated through extensive simulations, and the results demonstrate that LPMC performs better than the existing schemes in terms of packet delivery ratio, network throughput, end-to-end delays, and energy consumption.

The rest of the paper is organized as follows. In Section 2, we explain the existing multi-channel mechanisms for WSNs. We present the proposed mechanism in detail in Section 3. Section 4 demonstrates the performance evaluation of the LPMC. Finally, we conclude in Section 5 with a direction to the future works.

RELATED WORKS

In the existing literature, a significant number of MAC protocols (such as [12–15]) have been proposed for multi-channel communications in wireless networks. However, most of these protocols are not suitable for WSNs, because they assume that the transceiver can operate on multiple frequencies simultaneously or that the nodes are equipped with multiple radios, and current sensor nodes with only a single half-duplex radio transceiver cannot satisfy those assumptions. The idea of multi-channel protocols in WSNs is not new. A number of MAC protocols have already been proposed for WSNs [7–9]. To achieve multi-channel diversity, most of these protocols assign different channels to the contending sender-receiver pairs. The receivers (or the senders) in a two-hop neighborhood are assigned different channels in order to avoid interference and to increase capacity. However, due to multi-hop communications in WSNs, the nodes need to receive and forward packets in different channels. Therefore, the nodes frequently switch channels and experience packet losses. The channel switching causes considerable delays and a high degree of synchronization. Furthermore, the nodes require a sophisticated scheduling mechanism in order to find the rendezvous time for the sender-receiver pair.

A dynamic channel allocation method is proposed in DM-MAC [10] that uses a control theory approach to dynamically allocate the channels to each sensor. Initially all the nodes communicate on the same channel and when a channel becomes overloaded, nodes migrate to new channels. More specifically, whenever a channel becomes overloaded, some of the nodes switch to other non-interfering channels. Nodes in DM-MAC measure the success rates of medium access. Once a node figures out that lot of messages are lost due to collisions and interference, and causes the success rate of the current channel to fall below a certain threshold, the node considers switching channels. In contrast, nodes return to the previous channel if the success rate increases. Therefore, in a lightly loaded condition, the nodes use a single channel. In high network load conditions, the nodes use multiple channels to increase the network capacity and to deliver the data. However, the main problem with that mechanism is that the nodes change channels independently. In multihop communications, the forwarding nodes might find that the next hop is using a different channel. Therefore, nodes in a single path might need to switch between channels. Channel switching causes delays, and the overall throughput might suffer. In addition, the neighboring nodes in a path require time synchronization, and it is a challenge to keep both the sender and receiver in the same channel.

To avoid channel switching, TMCP divides the network into a number of sink-rooted disjoint subtrees [11]. Nodes residing on different trees are assigned different channels. Each sub-tree uses an orthogonal channel, and, thus, the nodes do not require channel switching. Note that instead of assigning channels to the nodes like DM-MAC, TMCP assigns channel to the sub-trees which allows TMCP to work with a small number of channels. The goal of TMCP is to partition the network to experience minimum intra- and inter-tree interference. The inter-tree interference is eliminated by using orthogonal channels for different sub-trees. In contrast, network partitioning minimizes the intra-tree interference. Finally, a greedy heuristic is used to partition the network to replace the NP-complete partitioning problem.

However, TMCP has a heavy initialization phase that is required in order to partition the network. The tree partition does not consider the data rates of the nodes, so the sub-trees might have different loads. Furthermore, changes in the routes might require a reinitialization, which is too costly for WSNs. Moreover, if a set of nodes sends data at a very high rate, static channel allocation cannot deliver the data. Therefore, we propose a dynamic channel allocation method, in which the forwarding nodes do not require channel switching and channels are added or removed wherever and whenever necessary.

PROPOSED MECHANISM

LPMC Overview

To describe LPMC, we introduce the following notation and terminology. We define the base station (i.e., sink) as an entity that collects data from the sensor nodes (sources). We consider that the sensor network is mainly used for data collection. The data collection scheme builds a tree that connects the sink and the nodes. Each node forwards the data along the tree.

It is obvious that the use of multiple channels increases the network capacity. The base station needs to sink (receive) all of the data sent in different channels. We, therefore, assume that the sink is equipped with multiple transceivers, each of which works in a different channel. A single node can generate data for multiple concurrent applications. The data of a particular node uses a single path. We assume that there are K orthogonal channels available for the WSN. A detailed discussion on the number of effective channels that can be used in WSNs with the CC2420 radio chip can be found in [11].

Our proposed mechanism, LPMC, aims at utilizing minimum number of channels and channels are dynamically added if particular channels become overloaded. If a single channel is sufficient, then it uses one channel. If the generated data requires more capacity, channels are added. In contrast, when the traffic does not need additional channels, channels are removed so that the nodes eventually use only one channel. Whenever a channel is added, it is assigned to a set of nodes in the overloaded tree branch and thereafter, data flows in an entire path uses a single channel. Therefore, nodes in a single path do not need to switch channel to receive data from the upstream nodes, or to forward data to the downstream node.

An important design consideration for LPMC is to allocate channels based on the network load. Therefore, we need to effectively estimate the active load of the network and to allocate or deallocate channels accordingly. In generic WSNs, the sensor nodes only have a local view of the overall network behavior. They might effectively measure the traffic load in a neighborhood, but they are not well-positioned to perceive the overall network status [16,17]. In contrast, a sink has a more comprehensive view of the overall network performance, since it receives the data generated by all of the sources. Given this perspective, a sink can operate the channel management functionalities more efficiently than would be possible with a decentralized approach. We therefore keep the channel management functionality of LPMC at the sink, whereas sensor nodes are engaged only in changing the channels. The term channel switching refers to the interchange between channels by a node to receive and forward packets, whereas channel changing refers to the assignment of a new channel to a set of nodes.

- At the sink, LPMC has four distinct logical components:
- The network load detection (NLD) component observes the packet arrival rates and sending rates of the sources, and decides whether or not the network is overloaded.
- The channel allocation and deallocation (CAD) component adds or removes channels based on NLD's report about the network load.
- The path update (PU) component dynamically divides a group of nodes using one channel into two groups, and assigns a new channel to one of the groups if a single channel is unable to deliver the data of all of the nodes.
- The channel changing (CC) component sends an explicit message to the nodes to change linebreak their channel.

The design of LPMC does not depend on any features specific to a particular MAC layer, except for changing the operating channel. Link-level retransmission can improve the performance of LPMC, but it is not essential. We assume that the sensor nodes run a routing protocol that selects a path from each source to the sink. In the following sections, we describe the detailed design of LPMC.

Multi-channel Communications System

The basic idea of LPMC is that all nodes in a single path use the same channel. Nodes do not need to switch channels in order to receive and forward packets. Therefore, in a tree-structure, the channel used by a neighbor of the sink (one-hop away node) is to be used by all of the nodes that forward their data through this node. We refer to this node as the channel deciding node (CDN). Figure 1shows a typical WSN environment, in which a set of nodes sends data to the sink using a tree. There are four CDN nodes in the figure ($c_1 - c_4$). The nodes that forward their data through a CDN create atree-branch (TB), and all nodes within a TB use the same channel. When a channel assignment takes place, all of the nodes in a TB change channels. However, multiple TBs can use one channel. Figure 1shows four TBs ($tb_1 - tb_4$) rooted at four CDNs.

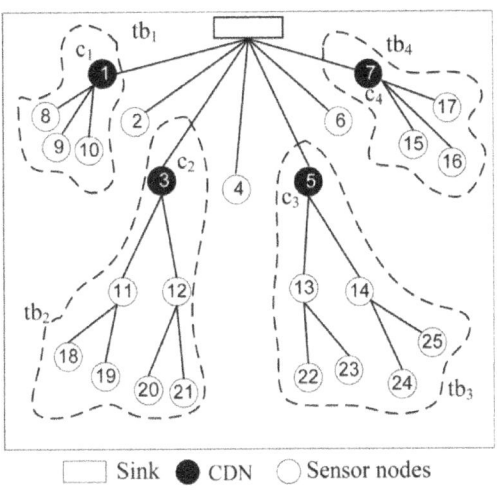

Figure 1: A typical WSN scenario with 4 TBs rooted at 4 CDNs.

When the network starts, all of the nodes use a predefined channel. We refer to this channel as theprimary channel. If the primary channel is overloaded, the most heavily loaded TB of the network is assigned a new channel. This continues as long as a channel is available and one of the allocated channels is overloaded. If there is no available channel, we assume that a rate control mechanism will restrict the data rates of the sources in order to avoid the packet losses. The rate control mechanism is beyond the scope of this paper. However, interested readers can refer to [16] and [18], where two well described rate control mechanisms for WSNs can be found. In contrast, when the network load decreases, the added channels are removed. If a single channel can handle the loads of two or more channels, the lowest channel ID is assigned to all of the nodes using these channels.

Network Load Detection (NLD)

One important technical challenge for LPMC is the design of a mechanism to estimate the active load or congestion level in the network or in a TB. Many techniques in the literature of wireless networks or wireless sensor networks measure the congestion level or load at a node. These techniques either measure the channel utilization around a node [19], the forwarding and reception ratio of a node [20], the buffer occupancy at the node [21,22], or a combination of both [18]. In contrast, LPMC estimates the network load from a different point of view. It assumes that the network is not loaded as long as the application's reliability is met. We define the reliability of a WSN application as the ratio of the number of packets received by the sink to the number of packets sent by the sources. Furthermore, LPMC aims at estimating the active load at the sink.

LPMC's load detection mechanism is based on the following intuition: a network (or a part of it) is not heavily loaded as long as the packet loss rate is acceptable, which permits packet losses due to a poor wireless link, medium contention, and transient congestion. When the network load increases, the packet loss rate also increases, or the interval between successive losses decreases. LPMC, therefore, uses the average loss interval as an active network load indicator.

The sink maintains a list of flows for each TB, and it maintains a per-flow list of missing packets and received packets based on the sequence number of the packets. The packets of a flow are forwarded in a single path, so the reception of an out-of-order packet indicates a packet loss. The sink also measures the number of successfully received packets before a loss event in order to measure the average loss interval. LPMC keeps track of the last n losses. Suppose, the sequence number of the packets of the m-th and the (m + 1)-th loss events of the i-th flow are s_m and s_{m+1}, respectively. Denoting $d_{i,m}$ as the length of the m-th loss interval of the i-th flow, we have $d_{i,m} = s_{m+1} - s_m$.

There are many techniques in the literature for measuring the average loss intervals. However, we choose the weighted average loss interval (WALI) method discussed in [23] over the others, because of its robustness in the parameters choices [16]. Therefore, for the last n losses, the average loss interval for flow i, denoted by $\hat{d}_i$, is calculated as

$$\hat{d}_i(1, n) = \frac{\sum_{m=1}^{n} d_{i,m} w_m}{\sum_{m=1}^{n} w_m},$$

$$\hat{d}_i(0, n - 1) = \frac{\sum_{m=0}^{n-1} d_{i,m} w_m}{\sum_{m=1}^{n} w_m},$$

$$\hat{d}_i = \max\left[\hat{d}_i(1, n), \hat{d}_i(0, n - 1)\right],$$

(1)

where $d_{i,0}$ is the number of successfully received packets since the most recent loss, and w_m is the weight assigned to the m-th loss interval. We have used n = 10, and $w_m = 1/m$ as our parameters. Furthermore, we assume that a smaller value of m indicates a recent loss interval, such that the parameters give greater weight to the recent loss intervals than to distant loss intervals. The reliability of the i-th flow, denoted as r_i, can be calculated as $r_i = 1 - 1/d'_i$. If r_i is less than the required reliability of the application, R_{req}, for any flow, then we say that the channel used by the i-th flow is overloaded.

However, if the application is loss intolerant (e.g., structural health monitoring application [24]) and lost packets are recovered by end-to-end retransmissions, then the required reliability is 1.0, and we cannot compare it with r_i. Therefore, we assume that the network is not overloaded as long as the loss rate is below a certain threshold. Furthermore, if a lost packet is recovered within the next n loss intervals, then we assume that the packet is not lost. Therefore, the sink maintains the history of the last n + q losses. If the m'-th lost packet is recovered by end-to-end loss recovery where $1 \leq m' \leq n$, the loss intervals of the last n losses are changed in the following way:

$$d_{i,m} = \begin{cases} d_{i,m}, & m < m' \\ d_{i,m} + d_{i,m+1}, & m = m' \\ d_{i,m+1}, & m' < m \leq n+1. \end{cases} \qquad (2)$$

If the loss rate of any flow exceeds the threshold, then we say that the channel used by the flow is overloaded.

The NLD also measures the active load of the network (i.e., the number of packets sent per unit time). The sink uses a timer for this. When the timer expires, it finds the sequence number of the most recently received packet of each flow and restarts the timer. If the two most recently recorded sequence numbers of the i-th flow are s_1 and s_2, then the number of packets sent for the flow is $S_i = s_2 - s_1$. The current load of a tree-branch, curr_load[tb], consisting of F flows is: $\sum Fi=1$ Si. The average instantaneous load, avg_load[tb], of a TB is measured by using the exponentially weighted moving average (EWMA) method as shown:

$$avg\_load[tb] = \alpha \times curr\_load[tb] + (1 - \alpha) \times avg\_load[tb], \qquad (3)$$

where α is a tuning parameter that is used to smooth the value of the average load of a TB. Through extensive simulation, we have set the value of α to 0.12 which produces the best estimation for a long-term average TB load.

Channel Allocation and Deallocation (CAD)

The channel allocation and deallocation component assigns the channels for the TBs. When an overloaded channel is used by multiple TBs, CAD assigns a lightly loaded or unused channel to one of the TBs. LPMC aims to use the minimum number of channels needed to satisfy the traffic load, so it first finds a lightly loaded channel that can be allocated to the overloaded TB (i.e., the TB with the highest loss rate). If it does not find such a channel, then an unused channel is allocated for the overloaded TB. However, if an overloaded channel is used by only one TB, then it cannot allocate another channel to the same TB, as in LPMC, all of the nodes in a TB should use the same channel. In this case, the path update component adds a new path by dividing an overloaded TB into two TBs, and assigns a new channel to one of the TBs.

In contrast, if the network becomes lightly loaded after having been at an overloaded status, then it removes one or more of the added channels. The idea of removing a channel is that the unused channel can be allocated to other nodes if necessary. A static channel allocation mechanism cannot do so. More specifically, if there is a continuous source of external interference for any channel, then the delivery ratio of the nodes that use this channel will be very low. A dynamic channel allocation mechanism can easily overcome this problem. First, because the sink keeps track of the achievable data rates of individual channels, it can determine if the loss of data packets is due to some reason other than overloading and can reallocate a separate channel. Second, even if the sink cannot compare it with the maximum achievable capacity, it adds a new channel for the nodes, which at least reduces the load of the channel. Therefore, LPMC tries to shrink the number of used channels. If two or more channels are shrink, the allocation mechanism keeps the channel with the smallest id as the active channel. The channels with higher id's are removed.

The CAD mechanism maintains two lists: the channel list with the fields < channel_id, status, max_load, curr_load and rem_load >, and the TB list with the fields < tb_id and avg_load >. The status of a channel is either used or unused. The CAD periodically obtains the average load (avg_load) of each TB from the NLD. The current load (curr_load) of a channel is the sum of the loads of the TBs using this channel. The maximum load (max_load) of a channel is the highest load that has been supported by the channel so far. When a channel is overloaded, the max_load is updated by the CAD; if the curr_load is higher than the max_load, curr_load becomes the max_load of the channel. The remaining load (rem_load) of a channel is the difference between the max_ load and curr_load.

Algorithm 1: Channel Allocation and Deallocation (CAD)

```
 1: Input: status[ ], max_load[ ], curr_load[ ],
 2:        rem_load[ ], avg_load[ ]
 3: ChannelAllocation (Channel i) {i-th channel is overloaded}
 4: Find the no of TBs, N, those use channel i.
 5: if N ≤ 1 then Call path update component and return.
 6: Find the TB with maximum loss rate, tb.
 7: for Each used channel j = 1 TO K, Except channel i do
 8:    if avg_load[tb] ≤ (1 − β) × rem_load[j] then
 9:        Assign channel j to tb and return.
10: end for
11: if unused channel available then assign it to tb.

12: ChannelDeallocation ()
13: for Each Channel i = 1 TO K do
14:    for Each Channel j = i + 1 to K do
15:        if curr_load[i] + curr_load[j] ≤ (1-β) × max_load[i] then
16:            Assign Channel i to the TBs using Channel j
17:            curr_load[i] += curr_load[j]
18:            status[j] = unused
19:        end if
20:    end for
21: end for
```

Algorithm 1 shows the detailed operation of CAD. The NLD component notifies the CAD about an overloaded channel. The channel allocation mechanism first checks the number of TBs that use the overloaded channel. If a single TB is using the overloaded channel, then CAD calls the path update component (see next sub-section). If there are multiple TBs, then CAD first tries to allocate a usedchannel; otherwise CAD allocates an unused channel if there are any. The channel allocation mechanism finds the TB that has the maximum loss rate. CAD tries to find a used channel with a remaining load that can accommodate the average load of the TB with the highest loss rate. However, the remaining load of a used channel is an estimated value, and an imprecise estimation can cause that channel (i.e., the channel which will be assigned) to be overloaded again. This might enforce another channel assignment. To avoid this, we have used a safeguard, β, which ensures that a certain percentage of the remaining load of a used channel is not considered when it is allocated. In the simulation, we have set the value of β to be 0.1.

In contrast, the channel deallocation mechanism removes channel when the load decreases. The NLD indicates when a channel is overloaded and CAD runs the ChannelAllocation function, whereas every time NLD updates the load of the TBs, CAD runs the greedy ChannelDeallocaton function. The ChannelDeallocation function checks whether a single channel (in addition to its current load) can accommodate the current load of another channel or not. If it finds such a channel, then, that channel is allocated the load of both the channels and the other channel is marked as unused.

The capacity of a channel might decrease for to many reasons, for example, very bad link quality, external interference, or even jamming by malicious nodes. LPMC changes the channel if the overall capacity of the channel is decreased to a certain fraction of the max_load of the channel. In this case, CAD assigns an unused channel for the TB(s). This feature of LPMC has an inherent benefit over static channel allocation schemes (i.e., TMCP) where it is not possible to dynamically measure the channel capacity and switch to an unused channel.

Path Update (PU)

In LPMC, a single TB uses only one channel, which ensures that a node does not need to switch channels to receive or forward packets. Therefore, whenever an overloaded channel is used by only one TB, CAD cannot allocate a new channel to that TB. A single TB using one channel can be overloaded due to many reasons, which include: (i) randomness of the node deployment, which place many nodes in a small area, (ii) dynamic path selection of the routing protocol, and (iii) nodes from a small portion of a large-scale dense network (usually far away from the sink) generating data at a very high rate. In such cases, LPMC partitions an overloaded TB into two TBs, and assigns a new channel to the newly created TB (i.e., a new TB consisting of some nodes of the overloaded TB). Therefore, paths are updated for a group of nodes in the overloaded TB, which will now use a new CDN to reach the sink. The path update component needs to find leaf nodes (i.e., a node that does not forward the data of other nodes) through which paths to the sink can be established. In case of failure to find such a node, we assume that a rate control mechanism is in place to restrict the source rates which will eventually decrease the channel load.

The path update module is initiated from the sink when the sink learns that an overloaded channel is used by a single TB. The sink first sends a unicast path update message (PUM) to the CDN of the overloaded TB. The PUM contains the following fields: < source, destination, type, tb_ID >. The sourceis the address of the PUM sender/forwarder, and destination is the address of an upstream node of the sender. LPMC assumes that every forwarding node keeps a list of its upstream nodes. The forwarding of a PUM is controlled by the value in type field. The type field has the value 1 or 2, and sink sets the type value to 1 while initiating a PUM. After receiving a PUM, a node either forwards it (when typevalue is 1) to its upstream node(s) or generates a PUM reply (when type value is 2). There may be cases where a PUM forwarder has multiple upstream nodes. In such cases, the PUM forwarder changes the type value to 2 and forwards the PUM to a randomly selected half of the upstream node(s). Because two or more branches join in this node, LPMC creates a new path for one of the branches and creates a new TB.

If a node receives a PUM with type value 2, it is forced to broadcast a path update message reply (PUMR) message. This PUMR creates a path from the node (i.e., PUMR generator) to the sink. The PUMR has the following fixed fields set by the source of the PUMR: < source, destination, and tb_ID >. The source is the address of PUMR generator, destination is the broadcast address, and tb_ID is copied from the PUM. Every PUMR forwarding node (including the PUMR generator) appends the following fields to the PUMR: < forward_node_addr, channel_ID, and hop >, where forward_node_addr is the address of the PUMR forwarder, channel_ID is its current channel, and hop is the hop count of the node from the sink.

Using the PUMR messages, the PUMR generator tries to find leaf nodes that can further forward the PUMR. However, such a node might be using a different channel. Therefore, PUMR is broadcasted in all of the channels one after another in order to find leaf nodes. The only leaf nodes that can forward a PUMR message are those with a hop count to the sink that is not greater than the value of the hopfield in the last entry of the PUMR. A node broadcasts the PUMR in a channel and hears the channel for some period. If no leaf node forwards the PUMR within this period, it broadcasts in another channel. Therefore, nodes first forward the PUMR in the receiving channel to ensure that the upstream node (previous broadcaster) hears it. Eventually, the PUMR is received by the sink. If the sink receives multiple PUMRs, it chooses the shortest path. The reverse path appended in the PUMR creates a new TB, and CC assigns a new channel.

We illustrate the path update operation with an example shown in Figure 2. Suppose that a channel is overloaded and is only being used by tb_3. The sink first sends a unicast PUM to node 5, which is the CDN of tb_3. Node 5 has two upstream nodes (i.e., 13 and 14), which means that two branches join in this node. It decides to create a new path for one of the branches. It randomly selects node 13 and sends a PUM with the type value set to 2. Therefore, node 13 becomes the PUMR generator and broadcasts PUMR in different channels to find leaf nodes with a hop count to the sink that is not greater than node 13's hop count to the sink. Node 4 is such a node, and the PUMR is eventually forwarded by node 4 to the sink. The reverse path appended in the PUMR creates the path from sink to the PUMR generator and creates a new TB. Node 4 becomes the CDN of the new TB, since it is one-hop away from the sink. The new paths from nodes 13, 22 and 23, now go through node 4 to reach the sink. Finally, the sink initiates a control message that enables all of the nodes in the new TB to change their channel. We discuss the channel changing procedure in the next sub-section.

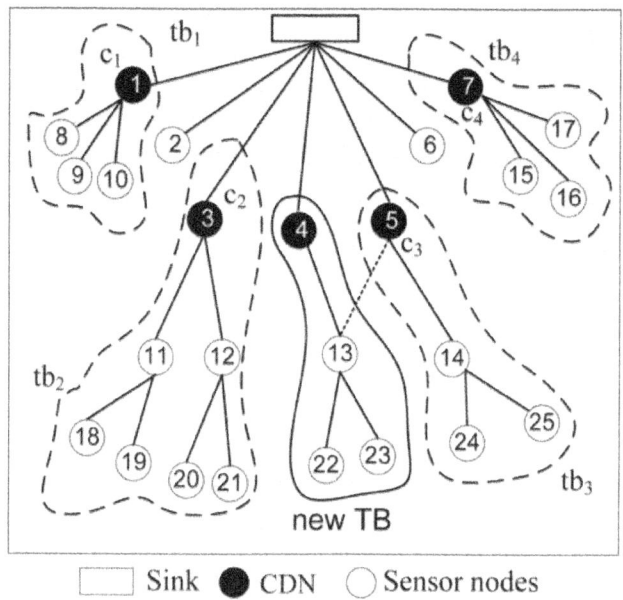

Figure 2: The path update module creates new paths for nodes 13, 22 and 23, by creating a new TB through node 4, and thus, divides the overloaded tb_3 of Figure 1 into two TBs.

Channel Changing (CC)

The channel changing (CC) component that resides in the sink issues explicit control messages that trigger all nodes in a overloaded TB to change their channel. Two types of channel changing messages (CCM) are used: (i) CCM-1 changes the channel of all nodes in a TB, and (ii) CCM-2 assigns a new channel in response to a path update.

To change the channel of an overloaded TB, the sink sends a unicast CCM-1 to the CDN of the TB. The message has the fields < sender, receiver, and new_channel >. All nodes broadcast the CCM-1 except the sink. After receiving a CCM-1, every node checks the sender field. If a node receives the CCM-1 from its downstream node, it forwards CCM-1; otherwise, it discards it. While forwarding the CCM-1, each node replaces the sender field with its own address and puts the broadcast address in thereceiver field. When a forwarder hears that at least one of its upstream nodes has forwarded the same message (by snooping), it changes its own channel to the new_channel. However, a leaf node changes to the new_channel after receiving the CCM-1 from its downstream node. The CCM-2 message is unicasted with the fields < sender, receiver, new_channel, and destination >. The destination is the address of the PUMR generator. CCM-2 also includes the reverse path from the sink to the destination, and the channel ID of each intermediate node. Every node forward the CCM-2, and changes its channel to the new_channel. When the destination receives the CCM-2, it converts it to CCM-1 and broadcasts it. Therefore, the path update and channel changing happen simultaneously.

Because the CCM-1 messages are broadcasted, the nodes that are missing this message need to find the channel ID of the downstream node. LPMC uses a What Is (WI) message to find the channel ID of a specific node. The node with the ID given in the WI message replies with a What Is Reply (WIR) message.

PERFORMANCE EVALUATIONS

Simulation Environment

We have performed extensive simulations to evaluate the performance of LPMC in NS-2 [25]. We have considered a network with an area of 200 m × 200 m and 250 nodes placed in a uniform random distribution. We have set the transmission power in such a way that the interference range becomes only 1.5 times the transmission range. In our experiment, the transmission and interference ranges of the nodes are set to 30 m and 45 m, respectively. Actually, this communication model is typically used to simulate the RF model of the CC2420 radio that operates on multiple-channels [11]. The link bandwidth is set to 250 kbps, and 6 orthogonal channels are used. Though the CC2420 radio chip used in Micaz motes provides 16 orthogonal channels, not all of them can be used in parallel because of close channel interference [26]. We use CSMA/CA as the MAC protocol with a maximum of 4 retransmissions. We compare the performance of LPMC with DM-MAC [10] and TMCP [11]. The required reliability (or success rate) is set to 0.95 for DM-MAC and LPMC. However, we set the required reliability of all the mechanisms to 1 when end-to-end reliability is considered. We have performed three different sets of experiments to evaluate the performance of the compared protocols. We show the impact of increasing the offered loads and varying the node densities in the first two set of experiments. Finally, we also show the impacts of the channel quality and external interference. All of the simulations were run 50 times, and the average results are plotted in the graphs. The other system parameters used in the simulation are summarized in Table 1.

Table 1: System parameters used in simulation

Parameter	Value	Parameter	Value
Link bit rate	250 Kbps	Packet size	32 Bytes
PHY Header	192 μs	MAC Header	224 bits
ACK Packet	112 bits	Slot Time	20 μs
SIFS	10 μs	DIFS	30 μs
Min CW	32	R_{req}	0.95
No. of channels	6	Switching delay	200 μs
α	0.12	β	0.10

Performance Metrics

We have considered the following performance metrics to evaluate the performance of LPMC in a different set of experiments: i. Network throughput-sum of the sizes of the total data packets received by the sink in a unit time, ii. Packet delivery ratio- the ratio of the total number of packets received by the sink to the number of packets sent by the sources, iii. Average end-to-end delay- the average end-to-end forwarding delay (which also includes medium access and switching delay) of the successfully delivered packets, and iv. Average energy consumption- the average energy consumed to successfully deliver a byte of data.

Simulation Results

Figures 3 and 4 show the performance comparison for 50 randomly selected sources. We have used the same set of sources for each mechanism. We have gradually increased the source data rates to measure the impact of increasing traffic load on the performance metrics. Figure 3(a) shows the network throughput with increasing data rates. When the source data rates are small (i.e., up to 6 packets/s, all of the mechanisms perform almost equally, because the network remains lightly loaded. However, as the offered load is increased, the channel capacities are exceeded and the performances of the respective protocols start to vary. Due to the static channel allocation, the throughput of TMCP depends on the locations of the sources. The channel with the highest number of nodes is overloaded, while many other channels remain underloaded. The dynamic channel allocation of DM-MAC achieves a higher throughput than TMCP until all of the channels in TMCP are overloaded. LPMC achieves the maximum throughput because it does not require channel switching and it allocates channels dynamically. Furthermore, the fair throughput of the sources (i.e., each source achieves the reliability) in LPMC is achieved up to a source rate of 11 packets/sec where the network throughput is 274 kbps. DM-MAC and TMCP achieve fair throughput up to source rates of 9 packets/s (where the network throughput is 216 kbps) and 7 packets/sec (where the network throughput is 174 kbps), respectively.

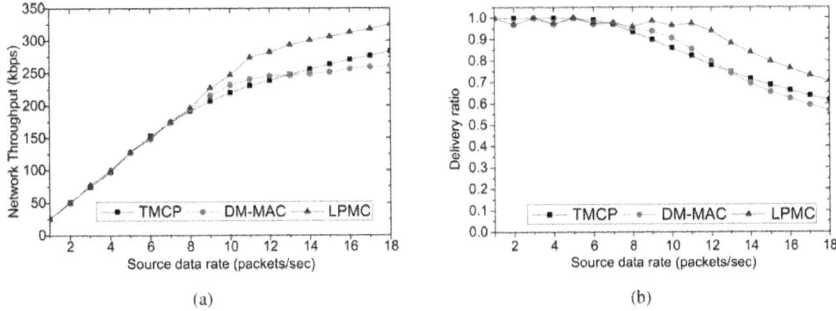

(a)   (b)

Figure 3: Performance comparison for randomly selected 50 sources with different data rates: (a) network throughput, (b) delivery ratio.

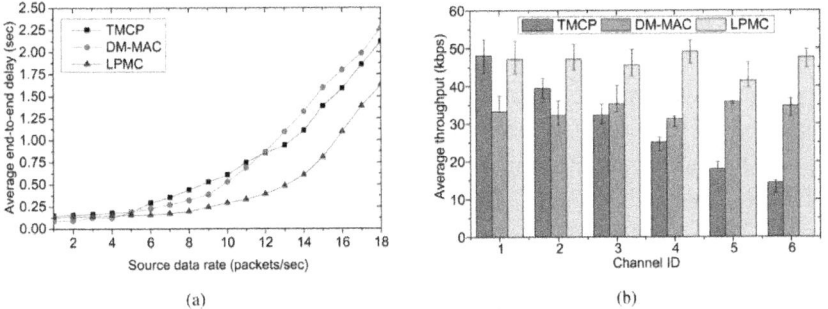

(a)   (b)

Figure 4: Performance comparison for randomly selected 50 sources with different data rates: (a) average end-to-end delay, (b) average channel capacity when all nodes achieve a fair throughput.

Figure 3(b) shows the delivery ratio at the sink, and as expected, LPMC outperforms the other two mechanisms at higher traffic loads. In LPMC, whenever a channel used by a single TB is overloaded, it updates the paths (i.e., it partitions the overloaded TB into two groups and creates a new TB that connects one group to the sink) and assigns a lightly loaded or unused channel to the new TB. Therefore, the number of packet losses due to congestion and contention decreases, and LPMC achieves a higher delivery ratio than TMCP and DM-MAC, while increasing the offered load. At higher traffic loads, DM-MAC experiences more intra-flow interference and therefore, DM-MAC's delivery ratio becomes less than TMCP's delivery ratio. Figure 4(a) shows the average end-to-end delays experienced by the successfully delivered packets. When the traffic load is low, all of the mechanisms experience small delays because the channels are operated with tolerable loads. Moreover, with a small amount of traffic, the channels do not become overloaded, which results in minimal channel switching (if any). However, as the traffic load is increased, LPMC outperforms both DM-MAC and TMCP in terms of the average end-to-end delay. This is because in LPMC, the nodes in a single path do not need to switch their channels to receive or forward a packet. TMCP has less delay than DM-MAC because of the static channel allocation. With DM-MAC, the delay is the highest, because the nodes in a single path need to switch channels more frequently. Figure 4(b) shows the channel utilization when the maximum fair throughput is achieved. TMCP achieves the minimum channel utilization, whereas it is at a maximum for LPMC. In case of TMCP, because of the static channel allocation, the channel utilization depends on the network topology. Therefore, the achieved throughput of different channels differs significantly. In contrast, DM-MAC and LPMC aim to use the smallest number of channels needed to satisfy the traffic load. As long as an acceptable delivery ratio is satisfied, LPMC does not add a new channel and achieves maximum channel utilization. Because of the added switching delay, DM-MAC cannot utilize the channel as fully as LPMC.

Figure 5 shows the results when end-to-end reliability is considered. In this case, the required reliability becomes 1 and lost packets are recovered by the end-to-end retransmissions. As shown in the Figure 5(a), with end-to-end reliability, the nodes using LPMC achieve a fair throughput up to a source rate of 7 packets/sec (where the network throughput is 176 kbps). Whereas, it is only 126 kbps for DM-MAC and 112 kbps for TMCP. Thereafter, all the mechanisms experience variations in achieved throughput because of the increased overhead caused by acknowledgments and end-to-end retransmissions. However, LPMC outperforms others as its load estimation policy (i.e., the NLD module discussed in Section 3.3) can effectively estimate whether a channel is overloaded or not, even when end-to-end reliability is considered. DM-MAC's load estimation only considers the local loss, and with the increased load of the end-to-end retransmission, most of the intermediate nodes are forced to change their channels (based on local condition), and they need to switch their channels for data forwarding. This increases the overhead, and decreases the network throughput. On the other hand, TMCP does not consider the load for its static channel allocation, and the channels become overloaded with the increased load of the end-to-end retransmissions. While LPMC can use the unused channels due to its dynamic channel allocation, TMPC's static allocation does not allow the nodes to change their channels. Especially, when the nodes send their data at a high rate, or number of nodes from a smaller area send their data simultaneously, which might overload a particular channel and the nodes using that channel have a lower throughput.

The dynamic channel allocation of LPMC can effectively divide the nodes into different channels and achieve a fair and efficient throughput, as long as the overall load is lower than the network capacity.

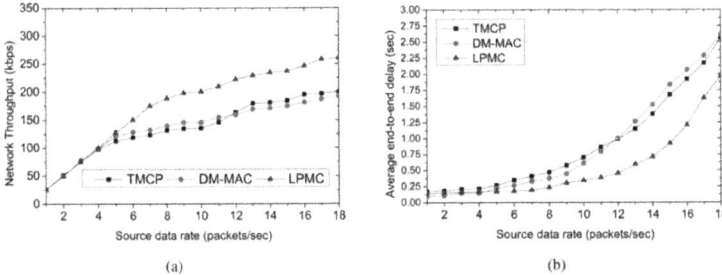

(a)   (b)

Figure 5: Performance comparisons for randomly selected 50 sources with different data rates when end-to-end reliability is considered (a) network throughput, (b) average end-to-end delay.

Figure 5(b) shows the average end-to-end delays incurred by the compared mechanisms as we increase the traffic load. We measure the average end-to-end delays of the successfully delivered packets. It is noticeable that average end-to-end delays of all the mechanisms with end-to-end reliability are higher than that of end-to-end delays (without considering end-to-end reliability as Shown in 4(a)). We measure that on an average only 7control overhead of the end-to-end retransmissions, overall network load increases which cause increased delay as compared to the absence of the end-to-end retransmissions. Initially, TMCP experiences higher end-to-end delays than that of DM-MAC and LPMC. Because of the static channel allocation of TMCP, the channels become overloaded fast and packets are lost due to increased congestion and contention, which increases the average end-to-end delays. On the other hand, when the traffic load becomes high, DM-MAC experiences more channel switching. This increased switching delay contributes to the increased end-to-end delay while using DM-MAC at higher traffic load.

Figures 6 and 7 show the performance comparison for a varying number of sources from a randomly selected location. Each source generates data at a rate of 20 packets/s. We measure the impact of increasing the number of sources on various performance metrics. Due to the static channel allocation, some of the channels are kept idle in TMCP, achieving the minimum throughput with more than 10 sources, as shown in Figure 6(a). DM-MAC requires channel switching when multiple channels are active, and the network throughput is less than that of LPMC. Figure 6(b) shows the delivery ratio of the different mechanisms, and LPMC achieves the highest delivery ratio. In the case of TMCP, some channels are overloaded quickly, while others remain underutilized, because the sources are selected randomly. Therefore, packets tend to get lost, due to the increased congestion and contention, and the delivery ratio drops significantly as the number of nodes is increased. In contrast, both DM-MAC and LPMC add channels when the network becomes overloaded and, thus, achieve a higher packet delivery ratio.

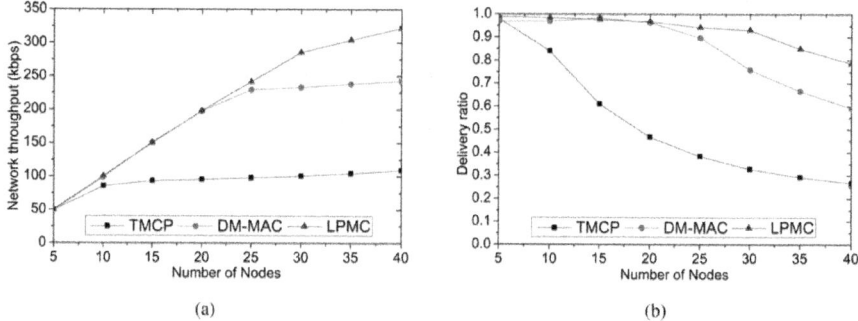

(a)   (b)

Figure 6: Performance with varying number of sources from a randomly selected location: (a) network throughput, (b) delivery ratio.

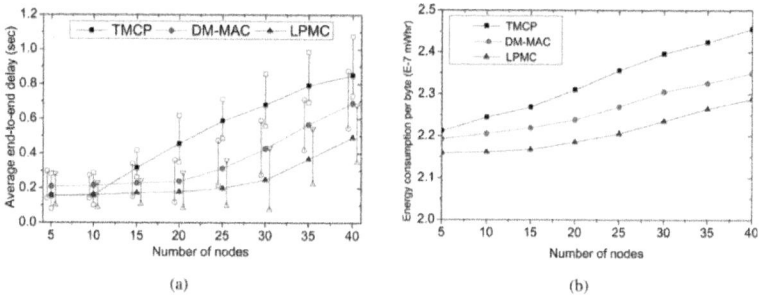

(a)   (b)

Figure 7: Performance with varying number of sources from a randomly selected location: (a) average end-to-end delay, (b) energy consumption per delivered data byte.

Figure 7(a) shows the average end-to-end delays of the packets. Because TMCP does not utilize all of the channels, the network becomes overloaded when the number of nodes increases. This increases the packet loss rate and more packets tend to be retransmitted, which increases the average end-to-end delay. At higher traffic loads, DM-MAC adds new channels to share the traffic of an overloaded region. However, unlike LPMC, the nodes in DM-MAC need to frequently switch channels in a path. The channel switching increases the medium access time in DM-MAC, so the delay is higher than that of LPMC. Finally, Figure 7(b) shows the energy required for each mechanism to successfully deliver a data byte. Because of the increased packet delivery ratio, the energy consumption in LPMC is lower than that of DM-MAC and TMCP as we increase the number of nodes. Figures 8 and 9 demonstrate the impact of the channel quality and external interference. We vary the packet loss rate randomly from 10 to 80 percent. All of the nodes send their data to the sink. Both LPMC and DM-MAC assign a new channel when the channel quality degrades, whereas the nodes using the interfered (or low quality) channel cannot deliver their data in TMCP. More specifically, when the traffic load is low, LPMC and DM-MAC can find an unused or new channel to replace a low-quality channel. However, when the network becomes overloaded, these mechanisms cannot find an unused channel that can be added to the network. Therefore, LPMC and DM-MAC achieve the required delivery ratio, as shown in Figure 8(b) (and therefore, a higher throughput as shown in Figure 8(a)), as long as the offered load is less than or equal to the aggregate channel capacity. However, LPMC achieves higher throughput and reliability than DM-MAC, because it does not require channel switching. In contrast, the throughput in TMCP increases for higher source rates, because of the static channel assignment.

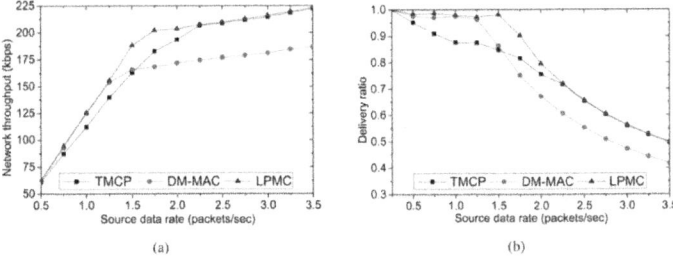

Figure 8: Impact of channel quality and external interference on the performance: (a) network throughput, (b) delivery ratio.

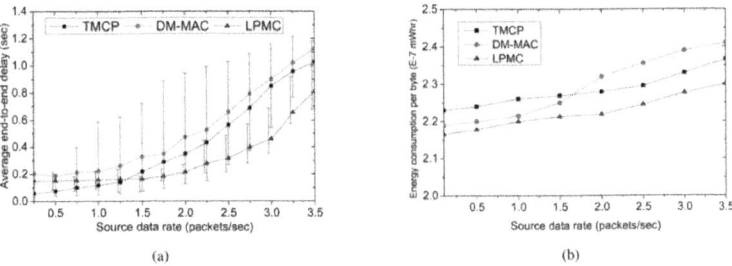

Figure 9: Impact of channel quality and external interference on the performance: (a) average end-to-end delay (The vertical lines shows the minimum and maximum delays for different channels), (b) energy consumption per delivered byte. All nodes send data to the sink.

Figure 9(a) shows the average end-to-end delays when the offered loads are increased. As the traffic load is increased, all of the mechanisms tend to experience more delays. However, DM-MAC has the largest delay, due to the channel switching required by the nodes to forward or receive packets. Because all of the channels are active in TMCP, the delay is lower than that of LPMC for low data rates. However, as the traffic load increases, the nodes that are using low-quality channels experience a higher delay in TMCP. Furthermore, TMCP has a higher delay than LPMC for higher source rates, because the queuing and medium access delay is very high for low-quality channels. The vertical lines show the maximum and minimum average delays among the channels.

Finally, Figure 9(b) represents the energy consumption per byte as we increase the load of the network. Since the packet delivery ratio of LPMC is more tolerant of an increasing network load, it also proves to be more energy-efficient than DM-MAC and TMCP. As shown in the figure, energy consumption for each successfully delivered data byte in LPMC increases from 2.23×10^{-7} mWhr to only 2.36×10^{-7} mWhr as we increase the network load.

CONCLUSIONS

In this paper, we have designed a load-adaptive multi-channel communications system for WSNs. LPMC controls the channel allocation and deallocation at the sink and dynamically adds or removes channels. The dynamic channel allocation utilizes the limited channels of WSNs very efficiently and adds channel(s) only when necessary. LPMC increases the channel utilization and network throughput, and reduces the delay. LPMC works very efficiently in a dynamic environment in which traffic pattern changes very frequently, wireless environment is very dynamic, and even the presence of external interference might destroy the communications of one or more channels. LPMC dynamically divides a branch of the networks into smaller branches if the load of the branch is more than the capacity of a single channel. Finally, the simulation results demonstrate that LPMC outperforms the existing mechanisms.

In the future, we would like to test the performance of LPMC in a real testbed. Another future interest lies in designing a multi-channel communications scheme for large-scale WSNs where multiple sinks are considered.

ACKNOWLEDGMENTS

This work was supported by the IT R&D program of MKE/KEIT. (2009-S-014-01, On the development of Sensing based Emotive Service Mobile Handheld Devices). Choong Seon Hong is the corresponding author.

REFERENCES

1. Zhang, H; Arora, A; Choi, YR; Gouda, MG. Reliable bursty convergecast in wireless sensor networks. Comput. Commun 2007, 30, 2560–2576.

2. Kang, J; Zhang, Y; Nath, B. TARA: Topology-Aware Resource Adaptation to Alleviate Congestion in Sensor Networks. IEEE Trans. Parall. Distrib. Sys 2007, 18, 919–931.

3. Kang, J; Nath, B; Zhang, Y; Yu, S. Adaptive Resource Control Scheme to Alleviate Congestion in Sensor Networks. Proceedings of the 1st International Conference on Broadband Networks (BroadNets'04), San Jose, CA, USA, 25–29 October 2004.

4. XBOW MICA2 Mote Specifications, Available online: http://www.xbow.com (accessed on 05 July 2010).

5. Polastre, J; Szewczyk, R; Culler, D. Telos: Enabling Ultra-low Power Wireless Research. Proceedings of the 4th International Symposium on Information Processing in Sensor Networks, IPSN '05; IEEE Press: Piscataway, NJ, USA, 2005; p. 48.

6. Hill, J; Szewczyk, R; Woo, A; Hollar, S; Culler, D; Pister, K. System architecture directions for networked sensors. Proceedings of the 9th International Conference on Architectural Support for Programming Languages and Operating Systems, Cambridge, MA, USA, November 2000; pp. 93–104.

7. Zhou, G; Huang, C; Yan, T; He, T; Stankovic, JA; Abdelzaher, TF. MMSN: Multi-Frequency Media Access Control for Wireless Sensor Networks. Proceedings of the 27th IEEE International Conference on Computer Communications, (INFOCOM), Singapore, 31 October–1 November 2006; pp. 1–13.

8. Zhang, J; Zhou, G; Huang, C; Son, S; Stankovic, J. TMMAC: An Energy Efficient Multi-Channel MAC Protocol for Ad Hoc Networks. Proceedings of the IEEE International Conference on Ccommunication, IEEE ICC '07, Glasgow, UK, 24–28 June 2007; pp. 3554–3561.

9. Chen, X; Han, P; He, QS; Tu, S; Chen, ZL. A Multi-Channel MAC Protocol for Wireless Sensor Networks. Proceedings of the 6th IEEE International Conference on Computer and Information Technology, CIT '06, Seoul, Korea, 20–22 September 2006.

10. Le, HK; Henriksson, D; Abdelzaher, T. A Practical Multi-channel Media Access Control Protocol for Wireless Sensor Networks. Proceedings of the 4th International Symposium on Information Processing in Sensor Networks, ACM/IEE IPSN '08, St Louis, MO, USA, 22–24 April 2008;

pp. 70–81.

11. Wu, Y; Stankovic, J; He, T; Lin, S. Realistic and Efficient Multi-Channel Communications in Wireless Sensor Networks. Proceedings of the 27th IEEE International Conference on Computer Communications, INFOCOM, Phoenix, AZ, USA, 13–18 April 2008; pp. 1193–1201.

12. Raniwala, A; Chiueh, T. Architecture and Algorithms for an IEEE 802.11-based Multi-channel Wireless Mesh Network. Proceedings of the 24th IEEE Annual Joint Conference of the IEEE Computer and Communications Societies, Pasadena, CA, USA, 13–17 March 2005; 3, pp. 2223–2234.

13. Bi, Y; Liu, KH; Cai, L; Shen, X; Zhao, H. A multi-channel token ring protocol for QoS provisioning in inter-vehicle communications. IEEE Trans. Wirel. Commun 2009, 8, 5621–5631.

14. Vedantham, R; Kakumanu, E; Lakshmanan, S; Sivakumar, R. Component based channel assignment in single radio, multi-channel ad hoc networks. Proceedings of the 12th Annual International Conference on Mobile Computing and Networking (MobiCom), Los Angeles, CA, USA, 23–29 September 2006; pp. 378–389.

15. So, J; Vaidya, N. Multi-Channel MAC for Ad Hoc Networks: Handling Multi-Channel Hidden Terminals Using a Single Transceiver. Proceedings the ACM International Symposium on Mobile Ad Hoc Networking and Computing, ACM MobiHoc, Tokyo, Japan, May 2004; pp. 222–233.

16. Paek, J; Govindan, R. RCRT: Rate-controlled Reliable Transport for Wireless Sensor Networks. Proceedings of the ACM Conference on Embedded Networked Sensor Systems (SenSys), Sydney, Australia, 6–9 November 2007; pp. 305–319.

17. Gungor, V; Akan, O; Akyildiz, I. A Real-Time and Reliable Transport (RT) protocol for wireless sensor and actor networks. IEEE/ACM Trans. Netw 2008, 16, 359–370.

18. Alam, MM; Hong, CS. CRRT: Congestion-Aware and Rate-Controlled reliable transport in Wireless Sensor Networks. 2009, 184–199.

19. Wan, CY; Eisenman, SB; Campbell, AT. CODA: Congestion Detection and Avoidance in Sensor Networks. Proceedings of the ACM Conference on Embedded Networked Sensor Systems (SenSys), Los Angeles, CA, USA, 5–7 November 2003.

20. Wang, C; Li, B; Sohraby, K; Daneshmand, M; Hu, Y. Upstream congestion control in wireless sensor networks through cross-layer optimization. IEEE J. Sel. Area. Commun 2007, 25, 786–795.

21. Rangwala, S; Gummadi, R; Govindan, R; Psounis, K. Interference-aware fair rate control in wireless sensor networks. Proceedings of the Annual Conference of the Special Interest Group on Data Communication (SIGCOMM), Pisa, Italy, 11–15 September 2006.

22. Hull, B; Jamieson, K; Balakrishnan, H. Mitigating Congestion in Wireless Sensor Networks. Proceedings of the ACM Conference on Embedded Networked Sensor Systems (SenSys), Baltimore, MD, USA, 3–5 November 2004; pp. 134–147.

23. Floyd, S; Handley, M; Padhye, J; Widmer, J. Equation-based congestion control for unicast applications. Proceedings of the annual conference of the Special Interest Group on Data Communication (SIGCOMM) and the Association for Computing Machinery (ACM), Stockholm, Sweden, August–September 2000.

24. Chintalapudi, K; Paek, J; Gnawali, O; Fu, T; Dantu, K; Caffrey, J; Govindan, R; Johnson, E. Structural Damage Detection and Localization Using NETSHM. Proceedings of 6th International Symposium on Information Processing in Sensor Networks(IPSN), ACM/IEEE IPSN 2006, Los Angeles, CA, USA, 19–21 April 2006; pp. 475–482.

25. Information Sciences Institute. NS-2 Network Simulator 2003.

26. CC2420 24 GHz IEEE 802.15.4 / Zigbee-ready RF Transceiver, Avialable online: http://www.chipcon.com (accessed on 5 July 2010).

Chapter 6

VIRTUAL INDUCTION LOOPS BASED ON COOPERATIVE VEHICULAR COMMUNICATIONS

Marco Gramaglia [1,2], Carlos J. Bernardos [2], and Maria Calderon [2]

[1]Institute IMDEA Networks, Avenida del Mar Mediterraneo 22, 28918 Leganes (Madrid), Spain

[2]Department of Telematics Engineering, Universidad Carlos III de Madrid, Avda. Universidad, 30, 28911 Leganes (Madrid), Spain

ABSTRACT

Induction loop detectors have become the most utilized sensors in traffic management systems. The gathered traffic data is used to improve traffic efficiency (*i.e.*, warning users about congested areas or planning new infrastructures). Despite their usefulness, their deployment and maintenance costs are expensive. Vehicular networks are an emerging technology that can support novel strategies for ubiquitous and more cost-effective traffic data gathering. In this article, we propose and evaluate VIL (Virtual Induction Loop), a simple and lightweight traffic monitoring system based on cooperative vehicular communications. The proposed solution has been experimentally evaluated through simulation using real vehicular traces.

INTRODUCTION

Every day, while we commute from home to work or from home to school, we likely traverse several detectors located along the roadsides that record our transit. Traffic data collected by these fixed sensors is used by public transport authorities (*i.e.*, city/regional/state) to improve traffic efficiency by, for example, warning users about accidents or congested areas, or planning new infrastructures. All this collected traffic data is processed by a central unit, which may decide to inform drivers about a potential event of interest. The most common and reliable technology used to collect traffic data are induction loops [1]. These loops are embedded in roadways in a square formation that generates a magnetic field. Magnetic loops count the number of vehicles and collect some information for each vehicle traversing the loop, such as the instant of time, speed, lane and type of vehicle. This technology has been widely deployed all over the world in the last decades. However, the deployment and maintenance costs of the induction loops (ILs) are expensive [2].

Another existing monitoring technique is the use of video cameras. At the beginning cameras were only used for remote surveillance, but with the relatively recent improvements in image recognition and data analysis, video cameras are also currently being used to monitor road traffic load and state. Each vehicle is uniquely identified by its license plate number and then tracked over a defined stretch of the road. Although the use of image recognition tools can mimic the outcome obtained by induction loops [3], this still requires the deployment of specialized fixed-infrastructure (*i.e.*, camera posts). Moreover, video cameras might not properly operate at night or in severe weather conditions (e.g., fog, heavy rain or snow).

Vehicular networks based on short-range wireless communications are a new paradigm widely investigated nowadays to develop novel innovative Intelligent Transportation Systems (ITS) for safety and traffic efficiency. One-hop wireless communications among vehicles and, between vehicles and infrastructure nodes, enable the design of cooperative systems that can support novel decentralized strategies for ubiquitous and more cost-effective traffic data gathering.

In this context, this article proposes and evaluates VIL (Virtual Induction Loop), a simple traffic monitoring system based on cooperative vehicular communications. The basic idea behind VIL is to define virtual loops whose position is advertised by roadside units (RSUs) along the road. Vehicles, which are supposed to be equipped with Global Positioning System (GPS) receivers, monitor when they are traversing one of these virtual loops and store their state at that moment (e.g., time, speed and, lane). As soon as the vehicle gets close to an RSU, it delivers this recorded information to the RSU, which in turn collects information related to several vehicles and VILs and sends it to the central ITS station.

The resulting advantages from the deployment of VIL are manifold. More roads than those currently equipped with a monitoring infrastructure (mainly induction loops) could be easily observed without requiring additional costs (*i.e.*, nowadays only major urban cities can afford deploying and maintaining a monitoring infrastructure). VIL is a flexible and simple solution that incurs low communication overhead. VIL service is easily deployed using Context-Aware Messages (CAM) [4], standardized by the European Telecommunications Standards Institute (ETSI) to improve safety and traffic efficiency in roads. Furthermore, other foreseen ITS services may share the same communication infrastructure (e.g., pollution management).

The rest of this article is organized as follows. A brief review of ITS ETSI standardization is presented in Section 2. In Section 3 we detail our proposal, which is experimentally evaluated using a trace-driven simulator in Section 4, before concluding in Section 5.

BACKGROUND

The ETSI Technical Committee for Intelligent Transport Systems ETSI TC ITS [5] is currently developing a set of protocols and algorithms that define a harmonized communication system for European ITS applications. Different types of ITS stations (e.g., vehicles) are defined [6], which have the capability of communicating between them using different access technologies. In particular, the IEEE 802.11p [7] at the 5.9 GHz band, an amendment to the 802.11 protocol especially tailored for vehicular networking, is one of these access technologies.

Vehicles can communicate with each other or with fixed roadside ITS stations (also called Roadside Units, RSUs) installed along roads. The roadside units, which are likely to be deployed uniformly along roads (e.g., using SOS posts), are usually connected to a wired network infrastructure (e.g., the Internet) and have a wireless interface to communicate with vehicles. Through the continuous exchange of messages between vehicles (Vehicle-to-Vehicle or V2V communications), and between vehicles and infrastructure nodes (Vehicle-to-Infrastructure or V2I communications), real-time information about current road traffic conditions can be cooperatively collected and shared.

According to ETSI standards, ITS stations, vehicles and RSUs periodically broadcast secure Cooperative Awareness Messages (CAM) [4] to neighboring ITS stations that are located within a single hop distance. CAMs are distributed using 802.11p and provide information of presence, positions as well as basic state of communicating ITS stations (e.g., current acceleration, occupancy of the vehicle, current heading of the vehicle, ...). The periodic exchange of CAMs helps ITS stations to support higher layer protocols and cooperative applications, including road safety and traffic efficiency applications.

VIL OPERATION

A virtual induction loop (VIL) is a virtual line playing the same role as a legacy magnetic induction loop (IL). In this way, VIL service gathers real time information of the vehicles traversing this virtual line.

VIL is a traffic efficiency service that makes use of existing secure CAM messages sent by RSUs and vehicles [4]. The full operation is shown in Figure 1. First, the ingress RSU announces in its CAM messages the positions of the virtual induction loops present in the stretch of road under its influence. These CAM messages also include information on the identities of the egress RSUs, that is, the nearest RSUs the vehicle may find in its way, after traversing the virtual loops in this stretch, depending on the vehicle's trajectory (see Figure 1(a)). It is important to highlight two details: first, each stretch can not only have several egress RSUs, but also several ingress RSUs (*i.e.*, announcing the same virtual loops); second, an RSU may play the role of egress RSUs for a stretch and the role of ingress RSUs for the next stretch.

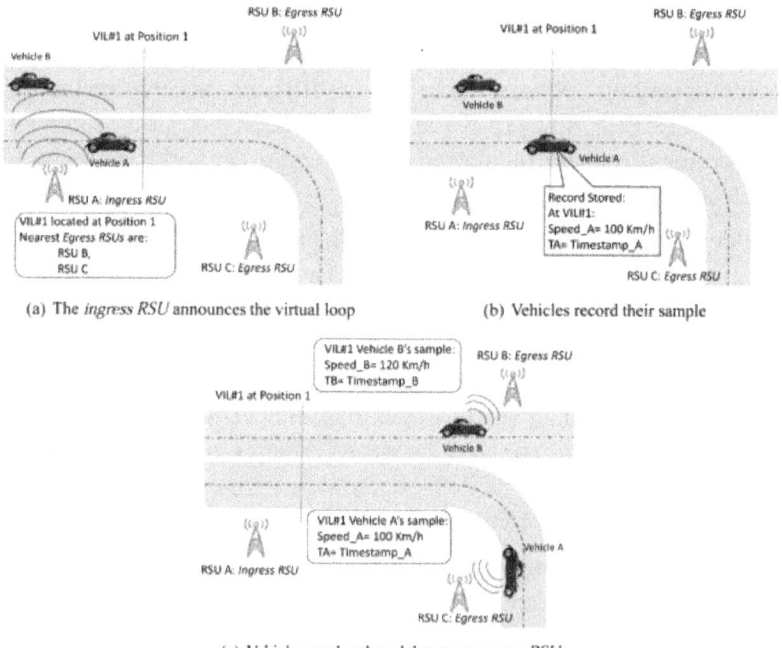

Figure 1: VIL operation.

On the other hand, GPS devices are now cheap enough to be included in vehicles configurations, as part of the on-board computer system. They supply two valuable pieces of information: the current position and the current time in Universal Time Coordinated (UTC) form. From these two variables, it is straightforward to calculate the current speed as well.

Thus, for each announced virtual loop a vehicle encounters during its transit within the monitored stretch (see Figure 1(b)), it records its state when traversing the virtual loop (e.g., timestamp, speed, lane, *etc.*).

At some point, the vehicle becomes aware of being within the radio coverage of an egress *RSU* upon the reception of a CAM message broadcast by the egress *RSU* (*i.e.*, the CAM message includes the identity of this egress *RSU*). From that moment on, and while being in the RSU's coverage area, CAM messages sent by the vehicle also include the information gathered when it traversed each of the virtual loops in the last stretch (see Figure 1(c)). In addition to the basic information (e.g., timestamp and speed), the vehicles can also upload another useful data such as the lane or their characteristics (e.g., traffic control centers often want to know the percentage of heavy lorries). The *egress RSUs*send the data gathered to the traffic control center. Once there, the information on traffic conditions is elaborated and eventually redistributed to drivers.

Impact of Technology Penetration Rate

Section 4 provides an evaluation of our solution, showing no significant differences between the data collected by using real induction loops and the one obtained from our virtual induction loop solution. So far, we have presented our solution assuming an ideal penetration rate situation, meaning that every vehicle on the road is equipped with a VIL-enabled ETSI TC ITS device. However, virtual induction loops can be deployed even in non-ideal scenarios in which not all the vehicles are VIL-enabled or have any communication capabilities.

If VIL is operated in a scenario where not all the vehicles participate in the virtual sensing, the estimated information will deviate from the real one. In order to correct this behavior, VIL can use a reference factor accounting for the ratio of vehicles that are VIL-enabled over the total of vehicles traversing a particular region. This can be easily achieved by using real induction loops that are already deployed to obtain reference values of the number of vehicles using a road, and then crossing that information with the number of VIL-enabled vehicles measured by a virtual induction loop placed at the same position of the real loop, as shown in Figure 2. Since the number of real induction loops is lower than the intended number of virtual loops, the reference value is obtained by averaging the information gathered from closest neighboring loops (e.g., loops located in the same city or the same metropolitan area). The time granularity used to estimate this reference factor can also be adjusted depending on the road characteristics. A similar correction factor, taking into account market penetration statistics, was also proposed by Bauza *et al.* in [8].

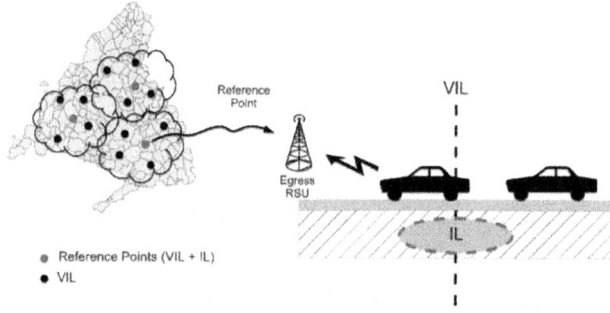

Figure 2: Reference factor estimation.

It is worth noting that even with lower penetration rates, VIL can provide an equivalent service to real induction loops, but as the number of VIL-equipped vehicles increases, VIL could also be used to enable more complex services, such as traffic congestion prediction.

EVALUATION

We evaluated our proposal by simulation. To achieve realistic results we built our software using the Veins framework [9] for OMNeT++ [10]. Veins [11] is an advanced vehicular communication simulator that couples realistic wireless features with the microscopic mobility simulator SUMO [12].

The goal of our simulator is to realistically investigate the effectiveness of VIL. To this aim, we consider both realistic wireless conditions and realistic vehicular patterns. To achieve the second goal we fed our simulator using real vehicular traces captured by magnetic induction loops. The data was kindly provided to us by the Madrid City Council and was collected along the M-30 orbital motorway (see Figure 3(a)). The trace is 2.5 hour long (from 8:00 AM to 10:30 AM) and it contains data of about 17,000 vehicles. A sample of the used data is shown in Table 1.

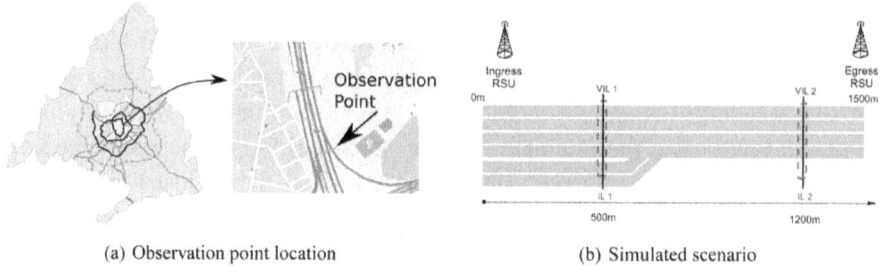

(a) Observation point location   (b) Simulated scenario

Figure 3: Real data and simulated environment.

Table 1: Trace sample

Timestamp	Vehicle #	Lane #	Speed (km/h)
08:57:00:6	5831	4	59
08:57:00:8	5832	6	61
08:57:00:9	5833	3	49
08:57:01:2	5834	4	68
08:57:01:7	5835	6	61
08:57:01:8	5836	1	68
08:57:01:8	5837	2	48

In our experiments the refresh frequency of the GPS is $f_r = 1Hz$ (update frequency of position and current time), which is an additional source of error in the experiments. However, this refresh frequency is considered realistic taking into account the features of the current commercial GPS devices.

Aiming at introducing realistic noise to the experiment, we introduce a Gaussian white noise for the two position components (both latitude and longitude) with a parameter $= 4m$, which is a fair error assumption for highways under clear conditions. The main simulation parameters are summarized in Table 2.

Table 2: Simulation settings

Simulation framework	OMNeT++, Veins and SUMO
Wireless Device	802.11g @ 6 Mb/s
Channel Model	Pathloss with channel fading
Monitored stretch length (m)	1500
VIL positions (from the *ingress RSU*) (m)	500, 1000
CAM frequency (s)	uniform (0.75,1.25)

The monitored stretch is 1,500 m long and it is a 6-lane single carriageway road, composed by 4 main lanes and 2 acceleration lanes. In the stretch there are one ingress *RSUs*, one egress *RSUs* and two virtual loops (see Figure 3(b)).

For the first 500 m, where the first VIL is placed, vehicles› speed is forced to be close (±2.5 m/s) to the departure one, then vehicles are allowed to increase their speed, up to 130 km/h. Figures 4, 5 and6 depict the obtained simulation results. As it is typically done, the vehicular metrics—flow and average speed—are summarized into coarser bins, in our case one-minute sized. In Figures 4 and 5, we compare the flow and the mean speed obtained by VIL at the egress RSU with the real ones from the vehicular traces obtained with real induction loops. It can be noticed that despite inaccuracy in latitude and longitude, and the errors introduced by the refresh frequency, the gap between the two curves is almost negligible. Figure 6 shows the cumulative distribution function (CDF) of the difference (in absolute value) between the real crossing timestamps and the ones measured by the VIL system. As it can be observed, almost all the crossing timestamps obtained by VIL fall into ±0.8 s of difference with the real ones

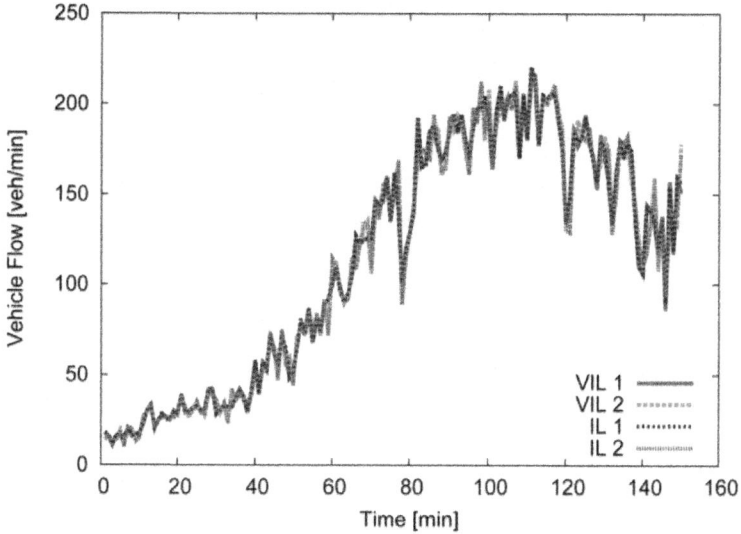

Figure 4: Vehicular flow.

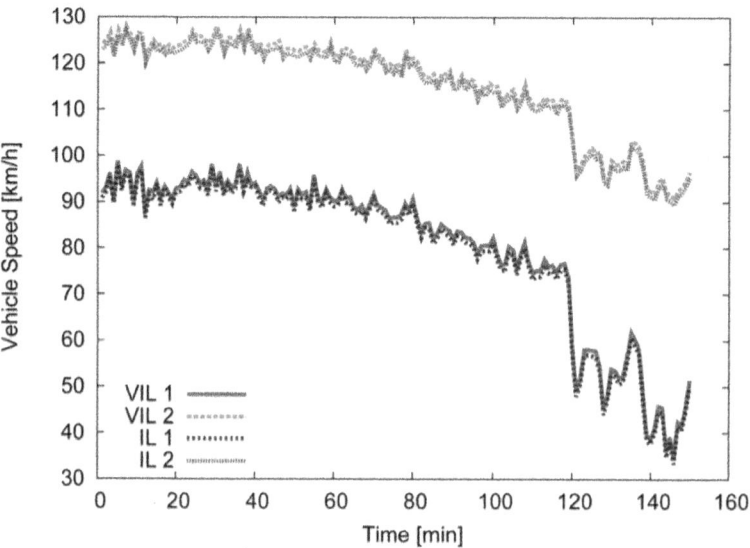

Figure 5: Average speed.

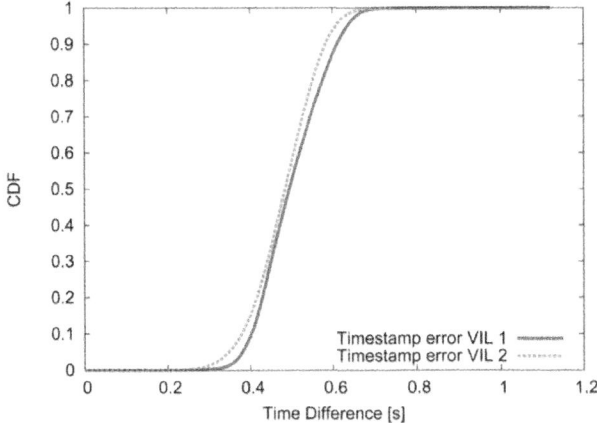

Figure 6: Crossing timestamps error.

A relevant parameter to be taken into account is the control overhead that the system introduces over the air. Due to the use of existing CAMs, no new messages are introduced by VIL system. However, the size of CAM messages is increased with the VIL information. In particular, CAMs broadcast by ingress RSU include the location of the virtual loops in the area under their influence (8 bytes per virtual loop) and the identity of the nearest egress RSUs (8 bytes per RSU). Considering the case of a CAM message that announces 3 virtual loops and 5 egress RSUs, the total increase in the CAM message size is 64 bytes. As for the CAM messages sent by the vehicles when they are in the coverage area of an egress RSU, the size of these CAMs is increased by 6 bytes (the timestamp field accounts for 4 additional bytes, and the speed one for 2 more bytes) for each virtual loop the vehicle has crossed. Therefore, the increase in the CAM message size can be considered to be of negligible impact.

CONCLUSIONS

Keeping the traffic conditions monitored is a crucial task in traffic management systems. Traditional solutions, based on the use of magnetic induction loops, are widely deployed, even though they incur high installation and maintenance costs. In this article we propose a simple approach based on cooperative vehicular communications that allows scalable monitoring without incurring high installation costs. Simulation results show that both the introduced error compared with the legacy methodologies and the wireless overhead are negligible. Even in initial deployment scenarios, where not every vehicle is VIL-enabled, our solution can be used as a suitable replacement of legacy monitoring systems. Moreover, VIL enables a configurable level of granularity, providing a higher data resolution just when and where it is needed.

ACKNOWLEDGMENTS

This work has been partially funded by the Spanish MICINN through the QUARTET project (TIN2009-13992-C02-01) and the I-MOVING project (TEC2010-18907). The authors would like to acknowledge the Madrid city council for kindly providing us with the empirical traces used in this work.

REFERENCES

1.   Anderson, R.L. Electromagnetic loop vehicle detectors. IEEE Trans. Veh. Technol. 1970, 19, 23–30.

2.   Leduc, G. Road Traffic Data: Collection Methods and Applications; JRC 47967; European Commission: Seville, Spain, 2008.

3.   Viarani, E. Extraction of Traffic Information from Images at DEIS. Proceedings of the 10th International Conference on Image Analysis and Processing, Venice, Italy, 27–29 September 1999; pp. 1073–1076.

4.   Intelligent Transport Systems (ITS); Vehicular Communications; Basic Set of Applications; Part 2: Specification of Cooperative Awareness Basic Service; European Telecommunications Standards Institute: Sophia Antipolis Cedex, France, 2011.

5.   Terms of Reference for Technical Committee (TC) Intelligent Transport Systems (ITS). Available online: http://portal.etsi.org/its/its_tor. asp (accessed on 22 January 2013).

6.   Intelligent Transport Systems (ITS); Communications Architecture; European Telecommunications Standards Institute: Sophia Antipolis Cedex, France, 2009.

7.   Jiang, D.; Delgrossi, L. IEEE 802.11p: Towards an International Standard for Wireless Access in Vehicular Environments. Proceedings of the IEEE Vehicular Technology Conference, Singapore, 11–14 May 2008; pp. 2036–2040.

8.   Bauza, R.; Gozalvez, J. Traffic congestion detection in large-scale scenarios using vehicle-to-vehicle communications. J. Netw. Comput. Appl. 2012. in press.

9.   Veins–Vehicles in Network Simulation. Available online: http://veins. car2x.org/ (accessed on 22 January 2013).

10. OMNeT++. Available online: http://www.omnetpp.org/ (accessed on 22 January 2013).

11. Sommer, C.; German, R.; Dressler, F. Bidirectionally coupled network and road traffic simulation for improved IVC analysis. IEEE Trans. Mob. Comput. 2011, 10, 3–15.

12. SUMO Simulation of Urban MObility. Available online: http://sumo. sourceforge.net/ (accessed on 22 January 2013).

Chapter 7

A MULTIMEDIA DATA VISUALIZATION BASED ON AD HOCCOMMUNICATION NETWORKS AND ITS APPLICATION TO DISASTER MANAGEMENT

Youhei Kawamura [1], Markus Wagner [2], Hyongdoo Jang [3], Hajime Nobuhara [1], Takeshi Shibuya [1], Itaru Kitahara [1], Ashraf M Dewan [4] and Bert Veenendaal [4]

[1]Faculty of Engineering, Science and Information Systems, University of Tsukuba, Ibaraki Prefecture 305-8573, Japan

[2]School of Computer Science, University of Adelaide, Adelaide, SA 5005, Australia

[3]Department of Mining Engineering and Metallurgical Engineering, Curtin University, Perth, WA 6433, Australia

[4]Department of Spatial Sciences, Curtin University, Perth, WA 6433, Australia

ABSTRACT

After massive earthquakes and other large-scale disasters, existing communication infrastructure may become unavailable and, therefore, it can be quite difficult for relief organizations to fully grasp the impact of the disaster on the affected region. Consequently, this will be the cause of delays to offer the strategic assistance, and to provide water and food, *etc.* In order to solve the problem of re-establishing communication infrastructure to allow for information gathering, we developed an *ad hoc* mobile communications network for disaster-struck areas using ZigBee. As the communication speed of ZigBee is low, we propose a problem-specific image compression method for the multimedia data visualization. By using the proposed method combined with GPS information, it is possible to quickly grasp the damage situation in the region. Through our communication experiments in Tsukuba City, Japan we confirm the effectiveness of our system as a disaster information gathering and management system.

INTRODUCTION

When large-scale disasters strike, a quick damage assessment is very important. Especially in the early stages, because of an allocation limitation of available manpower, it will be a key task to rapidly develop a rescue strategy according to precise information. For example, immediately after the Great East Japan Earthquake in 2011, people could not use communication systems, such as mobile phones and the Internet, in disaster-struck areas. With this lack of communication systems, the decision-making process of aid allocation and appropriate rescue operation was delayed [1,2,3,4]. As a result, emergency provisions were not adequately delivered to some evacuation centers.

In this series of research, a disaster management system has been proposed by Ishii and Kawamura [5], which collects and manages disaster information in an integrated fashion at control centers *via ad hoc* networks in order to support rescue efforts, distributions of relief goods, and so on. Concretely, in case of the disaster, investigators or volunteers install ZigBee terminals (see Figure 1 for an example) that have been stored in the evacuation centers or at other safe storage places in areas with interrupted communications infrastructure immediately after the occurrence of a major disaster. After installation, an *ad hoc* network is constructed automatically. The locations of these terminals are determined by the radio wave propagation analysis and optimal placement method that was developed by Ishii and Kawamura [5]. Investigators who are investigating the damages with the help of PCs or PDAs can gather information and transfer these data to the control center via the created ZigBee network. Additionally, it is possible that affected members of the public join the efforts of the professional relief organizations and share their information using their own PCs or PDAs.

Figure 1: ZigBee module used in this study.

Through this long-term research project, a disaster information management system will be developed and the communication speed, its availability, and the terminal installation time between the control center and evacuation centers will be assessed. As the ZigBee communication network standard is not suitable for the transmission of large volumes of data such as image data due to its low transmission rate (256 kbps), we propose a new transmission method that is the combination of GPS information and improved PIC format compression that focuses on spatial information. In a real-world experiment, we assess the effectiveness and the compression ratios by using our proposed disaster information management system.

The structure of this article is as follows. In Chapter 2, we provide an overview of the whole proposed disaster information management system. Then, we propose our method of transmitting and receiving maps in low-speed communication systems in Chapter 3. Finally, we describe in Chapter 4 our experiments of sending and receiving data in Tsukuba City, Japan.

DISASTER INFORMATION MANAGEMENT SYSTEMS

Current Status of Disaster Information Management

Currently, Multi-Channel Access (MCA) wireless communication systems and Ushahidi are put into effect as disaster information management systems after disasters.

Ushahidi (Figure 2, Watanabe *et al.* [6,7]) is the system that was used when the riots occurred during the elections in Kenya (2008). This open source web service was created to share information where such incidents occurred. Concretely, users can send information, such as "There is a riot in the area A" or "There is crime in the area B" via email, which will then appear on top of Google Maps. It has also been used to help collect situational information about military actions in the Gaza Strip (2009), Haiti (2010), after the earthquake and tsunami in Japan (2011), to monitor the elections in India and Mexico (2009), and in other cases. When the Great East Japan Earthquake occurred, the disaster information managing system named "sinsai.info", which is based on Ushahidi, was not only used to grasp the situation of the disaster area, but also to collect and share information about water supplies, official announcements, notices, news, and evacuation calls. It was quickly available and lots of people could understand the situation with the help of the online information visualization. *Via*email, posts on websites, and *via* the information sharing service Twitter it was possible to collect information continuously.

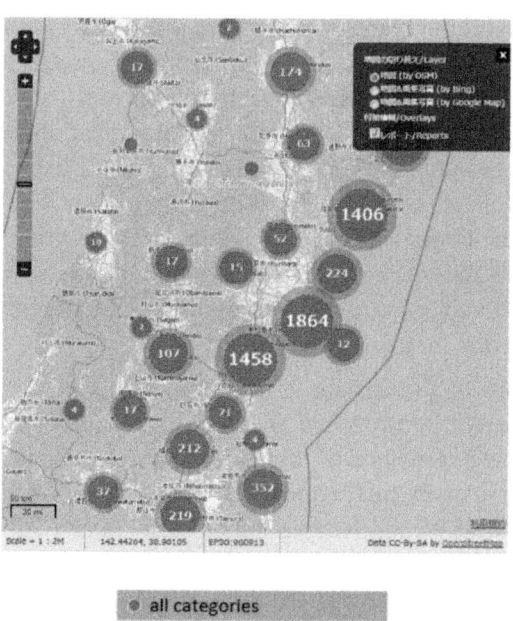

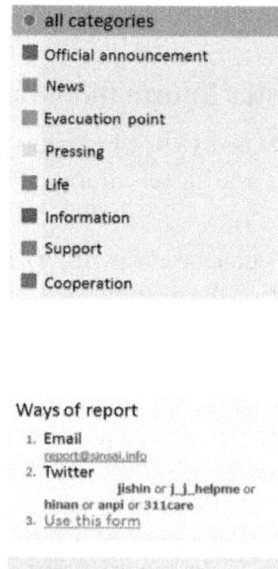

Figure 2: One example of a system based on Ushahidi.

In contrast to the web service Ushahidi, MCA (Multi-Channel Access System) is a commercial wireless communication system (Figure 3). It supports a large number of system users and can use a variety of radio channels. This system consists of a "control station" that the operating body installs and manages, and of "command" and "mobile stations" that the users install and manage in order to achieve radio communication between the units. For example, this system can be used by local governments to share information with fire-fighters and emergency personnel, even if both teams use different wireless communication technology.

MCA Control station

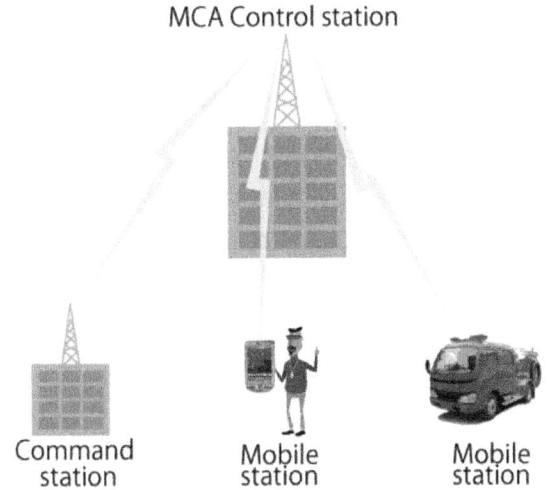

| Command station | Mobile station | Mobile station |

Figure 3: One example of a system based on the Multi-Channel Access System.

Overview of the Proposed Zigbee Disaster Control System

When a large-scale disaster such as the Great East Japan Earthquake occurs, the damage can be severe enough to take out MCA control stations, as well as fixed-line telephones, mobile phone stations, *etc.* Consequently, it can become difficult to use any of the two above-described disaster information management systems. We consider it an essential requirement for disaster control systems that wireless communication system can work even in such situations where hardly any previously existing infrastructure is available.

In this long-term research project, we develop a disaster information management system that uses the ZigBee communication standard (http://www.zigbee.org/). Figure 4 shows the overall flow of the proposed system. The disaster information management system is installed on PCs located in the control center and in the evacuation centers. This system contains the GIS data (*i.e.*, the geographical map data). The following is the organizational flow of the proposed system:

[Step ①] Immediately after the disaster (Corresponding to ① in Figure 4), the optimal allocation of ZigBee terminals between the control center and evacuation centers is calculated by using the PC in the control center.

[Step ②] The maps are updated with the of Step ① (Corresponding to ② in Figure 4).

[Step ③] Figure 5 illustrates a "ZigBee positioning situation". The communication terminal equipment carried by the investigators has a ZigBee communication module and a GPS module (corresponding to the red square in Figure 5); it must also display the coverage map using a PC or PDA (corresponding to ② in Figure 4). ZigBee terminals (assumed to have been stockpiled in the city hall or in evacuation centers within hours after the disaster) are placed along the way according to the computed optimal arrangement (Corresponding to ③ in Figure 4).

[Step ④] After the deployment of the ZigBee terminals, data transfers across the ZigBee network are available within the ranges that are visualized on the digital map (Corresponding to ④ of Figure 4). After this installation, this system will continue to work until the batteries run out after several months, or until the terminal are collected again.

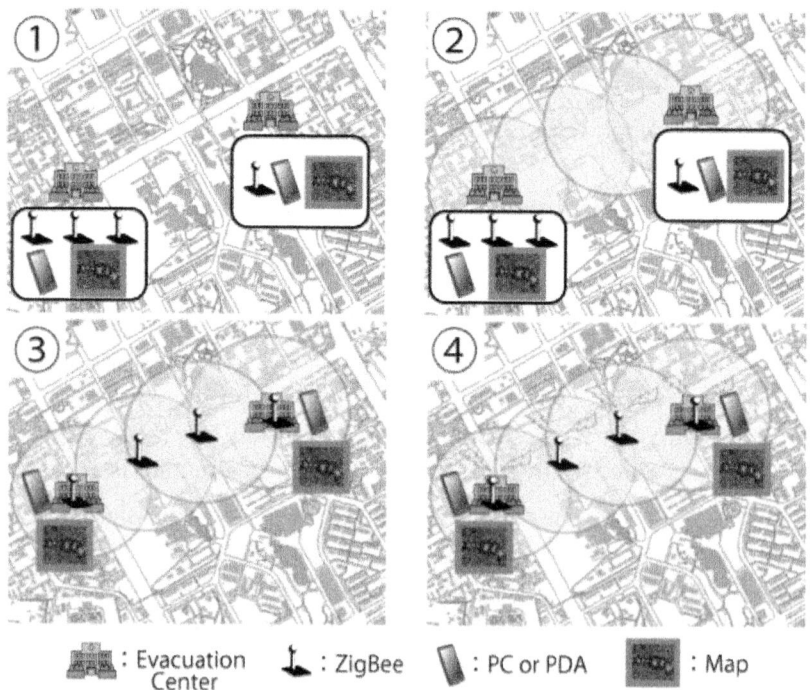

Figure 4: The process of establishing connections between centers: when the evacuation centers are established ①, the ZigBee network is planned ②, then the ZigBee nodes are located ③, and the network can be used ④.

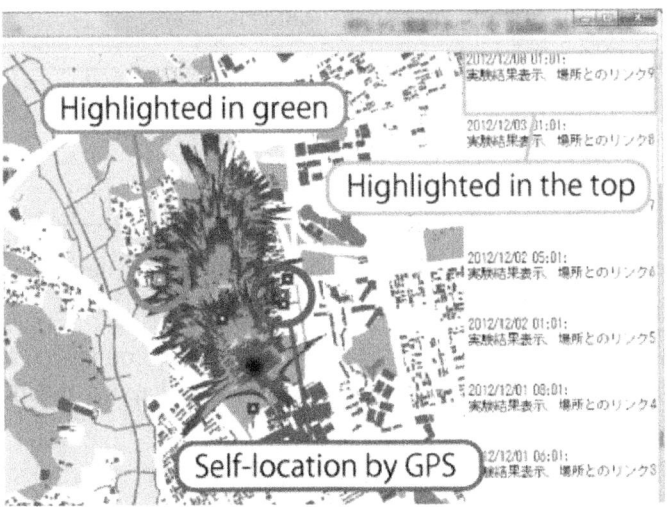

Figure 5: Screen of the proposed system.

Disaster information is information on a variety of materials, such as water supply information, official announcements, notices, news, and location of evacuation centers. In our case, disaster information consists of GPS data and text data, which is displayed on a map as a blue square. In our system, these texts (information) are displayed on the right of the map in chronological order, and once they are selected the corresponding information is highlighted. Additionally, GPS information is highlighted as a color change from blue to green (corresponding to the green of Figure 5), and the relevant text comes up to the first line (corresponding to the yellow of Figure 5). If the PDA or PC have been newly added to this system, or that system is re-connected after a temporary cut of the connection to the network, it is possible to acquire missing information by using the "request update function".

ZigBee Network

ZigBee is one of the communication standards for wireless sensor networks, and it has mainly been used in instrumentation control (see, for example, Kawamura *et al.* [2]).

Table 1 is a comparison of ZigBee with other typical wireless communication standards. Due to its characteristics, ZigBee is suitable, for example, for building automation, industrial automation, and home automation. Furthermore, ZigBee is also used in automatic meter reading, in security systems, and in remote control.

Table 1: Comparison of different communication standards

Name	ZigBee	Wi-Fi	Bluetooth
Communication distance	10–3000 m	100 m	10 m
Communication speed	250 kbps	11 Mbps	1 Mbps
Network capacity	65,536 nodes	32 nodes	7 nodes
Life-time on battery	several months	several hours	several days
Application	instrumentation control	wireless LAN	wireless accessories

ZigBee has the following four advantages for our scenario:

(1)   Support of a large network

As up to 65,635 terminals ("ZigBees") can be configured to form one wireless network, and the configuration can happen independently of location and time.

(2)   *Ad hoc* network

It is easy to build an *ad hoc* network, and more stable communication can be realized by providing a higher density of terminals along the intended transfer paths.

(3)   Low electric power consumption

Maintaining a long-term battery-powered network is possible.

(4)   Low cost

The costs of the ZigBee terminals are low, so it is possible to configure a wide area network by using a large number of them.

With the recent introduction of Bluetooth Low Energy (BLE) a competitor to ZigBee has been developed. While the energy consumption of BLE is significantly below that of Bluetooth, other features still make it unsuitable for our purposes (see for example, Mohan *et al.* [8] and Baker [9]). First, BLE is designed for small, local area networks and it is typically set up in star topologies or in one-to-one connections. ZigBee on the other hand is typically run in wider area networks and in mesh topologies that are significantly more robust when individual nodes fail. This is why we decided to build our disaster information management system based on ZigBee.

Despite these desirable advantages, ZigBee has the disadvantage of providing a relatively low communication speed. In practice this means that it is possible to send and receive data, such as text data of some kilobytes, however, it is difficult to transmit and receive multimedia data of some megabytes. To compensate for this problem, we are using a lightweight text-based communication in the present system. For the transmission of map data, we require significantly compressed data.

Optimized Arrangement Method and Radio Wave Propagation Analysis

It is necessary to construct the network as quickly as possible immediately after the disaster. In order to establish communication between the control center and all evacuation centers, we must decide on the installation locations of the ZigBee terminals. While deciding, we must consider the locations of woodland and water, because these influence the signal attenuation. There is also a need to respond flexibly to changes in the map information, as structures can collapse, and tsunamis and earthquakes can uproot forests. It is necessary to invest a small amount of computation time for estimating the radio wave propagation.

A radio wave propagation analysis method that was developed by Kawamura *et al.* is used in this system. This method uses a sequential tracking method that considers the transmission along the line-of-sight only (no transmission through obstacles, reflections, and diffractions) to reduce the computational complexity. The radio attenuation is estimated for different types of vegetation, and digital maps created based on GIS (Geographic Information System) to get the geographic information in the affected area (Ishii and Kawamura [5]). This method is then used by an iterated ray-launching method that greedily connects the control center with the evacuation sites; the eventual terminal locations are then determined along the paths of the simulated connections. The same algorithms are used by Moridi *et al.* [10] in a mine monitoring and communication system.

TRANSMITTING AND RECEIVING MAP-DATA IN LOW-SPEED COMMUNICATION

It can be expected that severe disasters change the geographical information, for example, the vegetation is changed, as are man-made structures. Nevertheless, it is necessary that members of the public and of relief organizations can communicate, even when the variations happen. In order to deal with such alterations, our system predicts the radio wave propagation and computes the optimal arrangement using the new map that is obtained and sent to each evacuation center after the disaster.Figure 6 shows the proposed map data transmission method. New map data based on satellite images can be available a few hours after a disaster. This, then, allows us to optimally arrange the terminals based on the situation at hand. Investigators, who have PCs or PDAs to collect disaster related information, need the map of the area and also a visualization of the available range of the radio communication. Therefore, an operator at an evacuation center needs to send updated geographical information (bit-mapped image, ① in Figure 6 shows in the red square the map as updated by a person who is present at the site) based on updated GPS information that is obtained via PCs and PDAs. All information will be combined for the modified map (② in Figure 6 shows the original map, ③ in Figure 6 shows the new map).

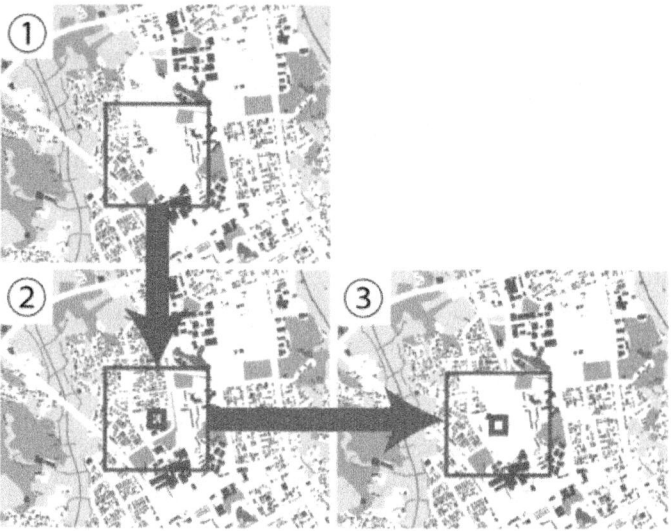

Figure 6: Map data update method.

Initially, this map image would be a bitmap and its image size would be too large to send across low-speed communication networks, such as a ZigBee network; therefore, image-compression is necessary. In this chapter, we propose a new problem-specific image-compression method.

Existing Image-Compression Methods

The PIC image format can be used for images that have continuous large areas of same colors, such as maps showing geographical information. Our method adopts PIC's run-length encoding method to longitudinal (vertical) direction and it has the characteristic of invertible transformation; for an in-depth description, see Carlson [11].

Figure 7 shows the image-compression method by PIC. Figure 7a shows the original state of the image. First of all, the picture is explored for each pixel from upper left to lower right line-by-line. If the color of an exploring pixel is same as some of the next pixel, the color information of the next one is deleted. If the color of the exploring pixel is different from some of the next pixel, the color information of next one is kept. This process is repeated until the lower right corner of the image is reached in order to find all the points where the color changes. Afterwards, all such points form "chains" from the top to the bottom of the image (Figure 7, ■ and ▲). In the actual compression step, the distances between the points where the colors change is recorded, as is the color itself, and also the direction of the chain and the offset.

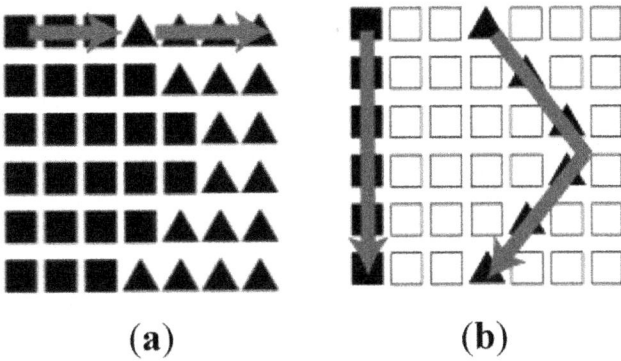

(a)   (b)

Figure 7: PIC format. (a) shows the original picture, (b) the right outlines the compression technique.

Proposed Image-Compression Method

The proposed method is an adaption of the PIC method to improve the compression for disaster-related information management systems. Images of maps have few colors because the colors represent just a small number of geographical information categories. The correspondences of geographical information and colors are stored in a database. Since this database does not change, the control center, the evacuation centers, and the investigators can share it in advance. In the database, geographical information and colors are encoded using three bits, e.g., "000 = road, 001 = water area,*etc.*". We call this suggested compression method GIM-format (Geographical Information Image). In case further data compression is necessary, we will limit the GIM images to only the sector around the current investigators (suggested image data size is 400 × 400 pixels). We will call this method GIMS-format (GIM Sector).

Comparison of Image-Compression Methods

Figure 8 shows map images that have four different levels of detail. First, we compress each image with GIM, GIMS, PIC, GIF, PNG, and JPEG using Adobe Illustrator; then we compare the resulting data-sizes. The original image data size is 28,473 kB (3006 × 2622 pixels, bitmap format). Note that, in the case of the GIMS-format, we report the mean of nine 400×400 pixels cut-outs from the original image.

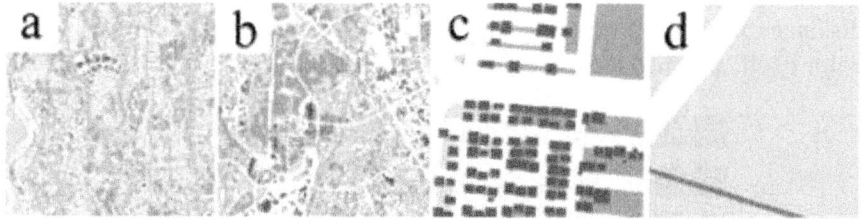

Figure 8: Image data used. The four figures (**a–d**) show shows map sections at four different levels of detail.

The results of the compressions are shown in Table 2. The GIM- and GIMS-format are the most effective compression methods. The resulting size of the GIM-format is approximately half of the others format images (except PIC-format) and is about 0.6 times the size of the PIC-format. In particular, the more large areas of the same colors are in the images, the more effective the compression is. PIC is more effective for images that have many small areas of same colors. This complements the effectiveness of the GIM-format for images that have many areas of same color.

Table 2: Comparison of compression methods, number are file sizes in KB

	a	b	c	d
GIMS	3.62	3.43	0.43	0.14
GIM	252	161	21	6
PIC	355	226	26	8
GIF	567	350	79	20
PNG	615	444	66	50
JPG	980	824	268	187

It is shown that all images can be compressed to 4k bytes or less by cutting out the important sections; thus, dividing the images makes it possible to transmit only necessary map information via the low-speed ZigBee network.

EVALUATION EXPERIMENT FOR THE PROPOSED SYS-TEM

Outline of the Evaluation Experiment

We conducted three experiments to evaluate the proposed disaster control system.

(1)   Following the system flow, the time is measured for the calculation of the optimal locations of four ZigBee terminals.

(2)   The experimenters transmit and receive disaster related information at 20 different spots. On each spot, the communication speed and the communication quality were measured.

(3)   Then, the experimenters obtain new map information about their vicinity from a control center using GPS information and update it. After that, the new range of the radio communication is calculated and visualized.

The investigated area for the installation of the ZigBee terminals is 800 × 500 meters around Kasuga 4-chome of Tsukuba City, Japan. Figure 9 shows the calculated optimal arrangement of ZigBee terminals. In the first experiment, four ZigBee terminals were placed; the time was measured while an experimenter placed the terminals.

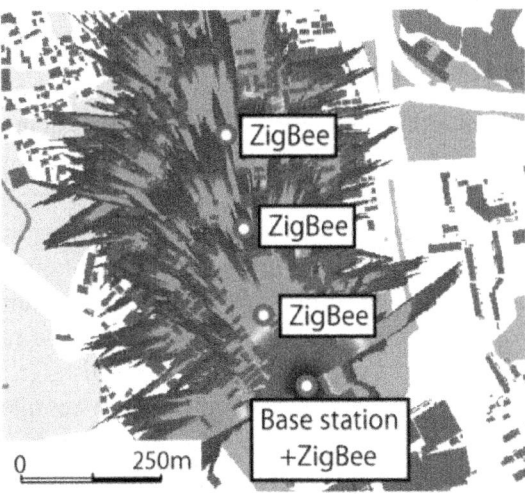

Figure 9: Figure of placement plan.

In the second experiment, the data transmitted and received consists of GPS data and "a communication experiment spot + the number of spot". The data was transmitted from a PC at an evacuation center to a PC on a control center through the ZigBee network. Additionally, the time required to send text data was measured using synchronized clocks on both PCs. By comparing the time-stamps of the messages with the receiving PC's time, we could calculate the transmission times. In addition, we consider a special (and likely) scenario in this second experiment: in one of the 20 measurements, an investigator leaves the area with ZigBee network coverage while investigating (either accidentally or on purpose) and then returns back into the ZigBee covered area. For this scenario, we have implemented a "request update function" to fetch data from the evacuation center was not received due to the missing signal.

In the third experiment, the map image only around an investigator (experimenter) was cut out and the map data was compressed at the evacuation center. The compressed image was sent to the control center and uncompressed. This process was conducted five times at different spots. As inSection 3, the original image size was 3006×2622 pixels, while the images were cut out into the size of 400×400 pixels for the GIMS-format.

Results of the Evaluation Experiment

In this section, we describe the results of our study in Tsukuba City, Japan. In this study, we measured the setup time needed, we measured the availability and speed of our ZigBee network, and we updated the geographical information locally and centrally.

Installation and Preparation Time

The experiment's starting time was 13:40 and system installation and preparation was completed at 14:12. In these roughly 30 min four ZigBee terminals (including the control center) were placed according to optimal arrangement calculated with a PC in an evacuation center. Based on results of a previous experiment in this area, it was assumed that one person can cover (investigate) 0.4 km^2 area in 30 min. We can use these numbers to estimate the resources needed in a real large-scale disaster: if the residential area of Tsukuba City (241.07 km^2) would be struck by a disaster, only 100 people (with 24 ZigBees each) would be needed to establish complete coverage with our proposed system within three hours.

Communication Availability

Figure 10 shows the results of transmitting and receiving the multimedia data at 20 spots. 19 spots were located inside of coverage of ZigBee communication and one was located "out of range" as described above. At 13 spots (Figure 10, numbers on white background) out of the 19 spots, data reception was possible. At the six others spots (Figure 10, numbers on red background), communication failed. There are buildings or elevated areas between these spots (Figure 10, numbers 8, 10, 12, 14, 16). The reasons of these failures are probably diffractions of the radio waves. As reflection, diffraction, and permeation of radio waves were not considered when the radio wave attenuation was estimated, the estimation error will be large in case there are a lot of buildings at that moment. Data obtained by an experimenter at spot

15 (Figure 10, yellow square) was stored when it was observed, and it was sent after the experimenter came inside the coverage of the ZigBee network. Subsequently, the stored data was received at the control center. This was a successful proof of the concept that an investigator can collect data while being out of the communication range and then share this data when coming into the communication range again.

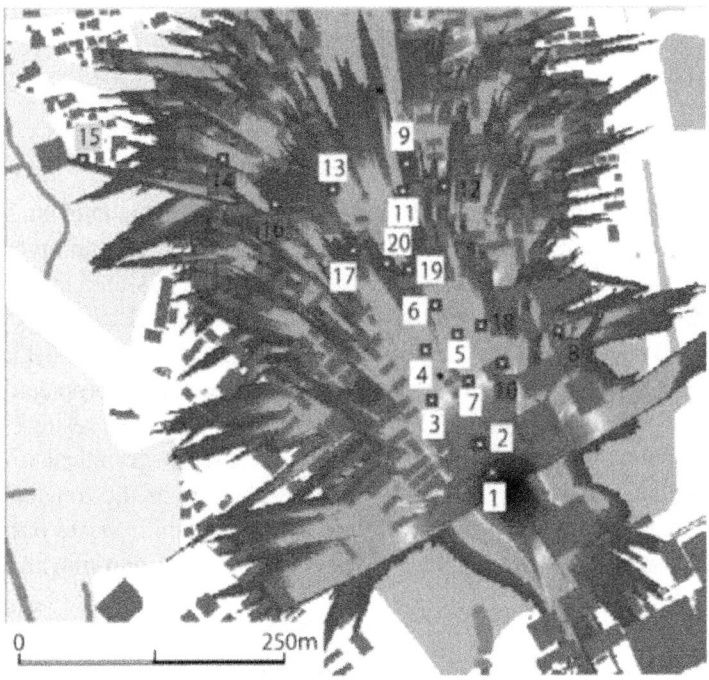

Figure 10: Experimental results of communication situation.

Communication Speed

Figure 11 shows the communication speed at each spot. Communication speeds were measured while the data was transmitted from an evacuation center to a control center at each spots. The images were transmitted and received at 300 ms intervals to prevent one image from splitting into two data sets. Many results concentrated around 300 ms regardless of the distance between the measured spot and the control center, and regardless of how many terminals the data went through. Interestingly, at three spots (Figure 11, numbers 5, 11, 17), results are around 700 ms. This means that first transmission or recipient failed, so the data was transmitted a second time. Thus, the resulting communication time increased to approximately 300×2 (ms).

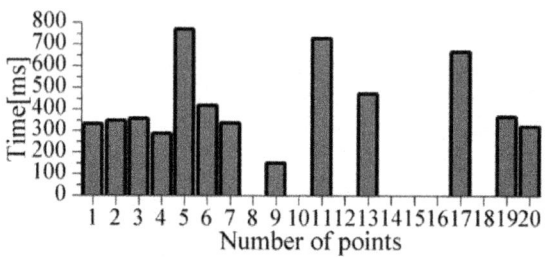

Figure 11: Experimental results of the communication speed tests.

Sending and Receiving Geographical Data Updates

Figure 12 shows the process of a geographical information update (image update) with a PC in an evacuation center. ① in Figure 12 is sent by an investigator to the control center after a disaster, and ② in Figure 12 is a stored image obtained before a disaster at an evacuation center. ③ in Figure 12 in the evacuation center is update from ② to ① only around the investigators GPS location. ④ in Figure 12 shows the result of prediction of radio wave propagation based on the old pre-disaster map and ⑤ in Figure 12 shows the result of it based on the updated post-disaster map. Using only the geographical information around an investigator significantly reduces the size of the transmitted data; moreover, the investigators know where to go to communicate based on the updated coverage prediction, so they can easily receive and transmit disaster-related information.

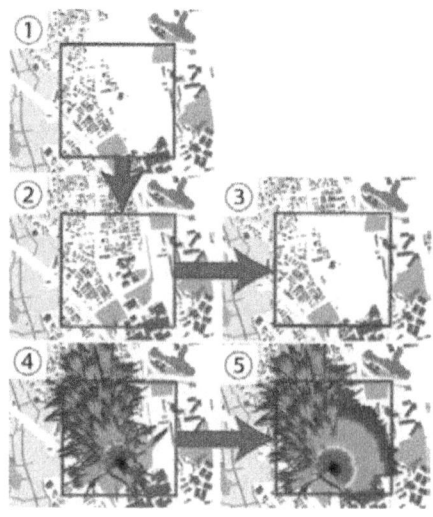

Figure 12: Experimental results of sending and receiving geographical data updates.

CONCLUSIONS

In this article, we described and evaluated a ZigBee wireless communication system that can be used to quickly gather information and to coordinate efforts when existing communication networks are become unavailable.

We proposed the GIM-format for geographical information and compared it with other image-compression methods. The size of GIM-format compressed images is approximately half of those generated by the other formats (except for the PIC-format). In particular, when the images have many large areas of identical colors (as it is the case of geographical data), the compression is very effective. In addition, we showed that the file size can be reduced further by limiting the image dimensions to the area of interest around the investigator's GPS location.

Through extensive experiments in Tsukuba City, we confirmed that our proposed disaster communication system can be used to gather disaster-related information and to send that data to the control center, which can then update the map for all investigators.

In addition, the experiment of transmitting and receiving map-data proved that the method is effective for the low-speed communication network of ZigBee terminals.

To conclude, the proposed disaster control system can be an alternative communication infrastructure when existing infrastructure is damaged by a major disaster. The system is easy to use even for laypeople, which enables the affected people to assist the professional organizations in their emergency relief efforts.

ACKNOWLEDGMENTS

The authors would like to thank the reviewers and the editors for their comprehensive feedback, which helped to improve the quality of the article.

AUTHOR CONTRIBUTIONS

All authors collaborated on this article in different ways that range from the inception of the idea, to the implementation of the tool, the conduct of the study, and the editing of the article. All authors contributed enough to warrant their coauthorship.

REFERENCES

1.   Sugit, H.; Fukuta, T.; Tamura, T.; Yokoi, T.; Hara, T.; Kashima, T.; Azuhata, T.; Shibasaki, B.; Yagi, Y. IISEE-NET: Information network

for earthquake disaster mitigation of developing countries. In Bulletin of the International Institute of Seismology and Earthquake Engineering; Building Research Institute: Ibaraki, Japan, 2003; pp. 137–143.

2. Kawamura, Y.; Dewan, A.; Veenendaal, B.; Shibuya, T.; Hayashi, M.; Kitahara, I.; Nobuhara, H.; Ishii, K. Using GIS to develop a mobile communications network for disaster-damaged areas. Int. J. Digit. Earth 2014, 7, 279–293.

3. Kawamura, Y.; Ishii, K.; Jang, H.; Wagner, M.; Nobuhara, H.; Dewan, A.; Veenendaal, B.; Kitahara, I. Analysis of radio wave propagation in an urban environment and its application to initial disaster response support. J. Dis. Res. 2015, 10, 655–666.

4. Nakahata, Y.; Kawamura, Y. Development of the landslide observation system using ZigBee. In Proceedings of the 2010 Society of Instrument and Control Engineers Annual Conference (SICE), Taipei, Taiwan, 18–21 August 2010; pp. 1191–1194.

5. Ishii, K.; Kawamura, Y. Analysis of radio wave propagation in an urban environment and its application to initial disaster response support. In Proceedings of the 2010 Society of Instrument and Control Engineers Annual Conference (SICE), Taipei, Taiwan, 18–21 August 2010; pp. 1195–1198.

6. Watanabe, N.; Uno, K. Utilization of sensor Web and spatial information platform as a supportive measure for field work monitoring. Rep. Inst. Sci. Tech. Res. 2011, 23, 62–66.

7. Watanabe, K.; Ichikawa, T. A study of the information exchanging system for a person requiring care at the time of earthquake disaster. SIG Tech. Rep. 2006, 92, 25–29.

8. Mohan, M.; George, N.; Davis, D. ZigBee technology for data communication—A comparative study with other wireless technologies. Int. J. Adv. Res. Comput. Sci. 2014, 5, 261–265.

9. Baker, N. ZigBee and Bluetooth strengths and weaknesses for industrial applications. Comput. Contr. Engin. J. 2005, 16, 20–25.

10. Moridi, M.A.; Kawamura, Y.; Sharifzadeh, M.; Chanda, E.K.; Wagner, M.; Jang, H.; Okawa, H. Development of underground mine monitoring and communication system integrated ZigBee and GIS. Int. J. Min. Sci. Technol. 2015. in press.

11. Carlson, W.E. A Survey of Computer Graphics Image Encoding and Storage Formats. Comput. Graph. 1991, 25, 67–75.

Chapter 8

DESIGN AND IMPLEMENTATION OF ELECTRONIC CONTROL TRAINER WITH PIC MICROCONTROLLER

Yousif I. Al Mashhadany

Electrical Engineering Department, Engineering College, University of Anbar, Baghdad, Iraq

ABSTRACT

This paper describes the implementation of a PIC microcontroller in a conventional laboratory-type electronic trainer. The work comprises software for the PIC and hardware for the software. The PIC controller uses an EasyPIC-6 board and includes a PC-interfaced programmer for the PIC chip. It has many external modules: 128×64 graphic LCD display, 2×16 LCD display, 4×4 keypad, and port expander, all in the same bench. The trainer is capable of 36 experiments in logic/analogue electronic and control systems. A 5-sided approximate sensor, two photoelectric sensors (BR56-DDT-P and BEN9M-TFR), four CMOS, four BCD-7-segment driven by CD4511B, two relays (2-pole and 3-pole), six voltages, ammeter measurement, DC motor, and 24VDC power supply, connect through connectors and pinions. Results of all the experiments show the trainer satisfying requirements of undergraduate and postgraduate projects involving conventional electronic and classical control systems.

INTRODUCTION

Modern microcontroller chips can store hundreds of thousands of transistors each. The first microprocessors had external peripherals such as memory, input-output lines, and timers (Matic, 2003). In time came a new device called integrated circuit (IC), which contains both processor and peripherals. Also called a microcontroller, this was the first chip with a microcomputer [1,2].

Peripheral Interface Controller (PIC) is new to electronics control. Providing complete control in a single chip, a PIC microcontroller has special function registers, power on reset, interrupts, user RAM for storing of program data, EPROM program memory, timer circuits, instruction set, low power consumption, and on-board Ato-D converters. It replaces conventional control of industrial machinery (e.g., motor-speed control) [2,3].

Microcontroller and microprocessor differ in many ways. In functionality, a microprocessor needs external components for receiving/sending data, and memory. A microcontroller does not need external components because all the necessary peripherals are built-in, saving time and space (seeFigure 1 for microcontroller set [4- 7]).

The EasyPIC-6 by MikroElektronika (see Figure 2) is an extraordinary development tool for programming and experimenting with PIC® microcontrollers. It supports over 160 MCUs in PIC10, PIC12, PIC16, and PIC18 families, in DIP packages from 8 to 40 pins. The board comes installed with PIC16F887. An impressive array of peripherals and expansion connectors are available onboard, as are optional LCD displays and temperature sensor [8,9].

An on-board programmer and mikroICD debugger allow direct connection to PC via USB cable. Fully functional demo versions of MikroElektronika's C, Pascal, and BASIC compilers are included (hex output limited to 2K program words), complete with documentation and dozens of sample programs. The EasyPIC-6 also includes an external ICD connector compatible with MPLAB ICD2 and ICD3, allowing full compatibility with MPLAB Integrated Development Environment (IDE) [10,11].

Its main problem is lack of facility for external experiments to be implemented in many undergraduate laboratory applications; it is also daunting to beginner designers. This paper presents a practical implementation of EasyPIC-6-based electronic control trainer able to execute about 36 experiments, and rearrangement of the EasyPIC-6 power supply to extend the trainer's capability to AC-DC-current applications.

Integrated Development Environment (IDE)

The core development tool set operates under the IDE umbrella called MPLAB. The tools look and feel the same,

Device	Program FLASH	Data Memory	Data EEPROM
PIC16F874	4K	192 Bytes	128 Bytes
PIC16F877	8K	368 Bytes	256 Bytes

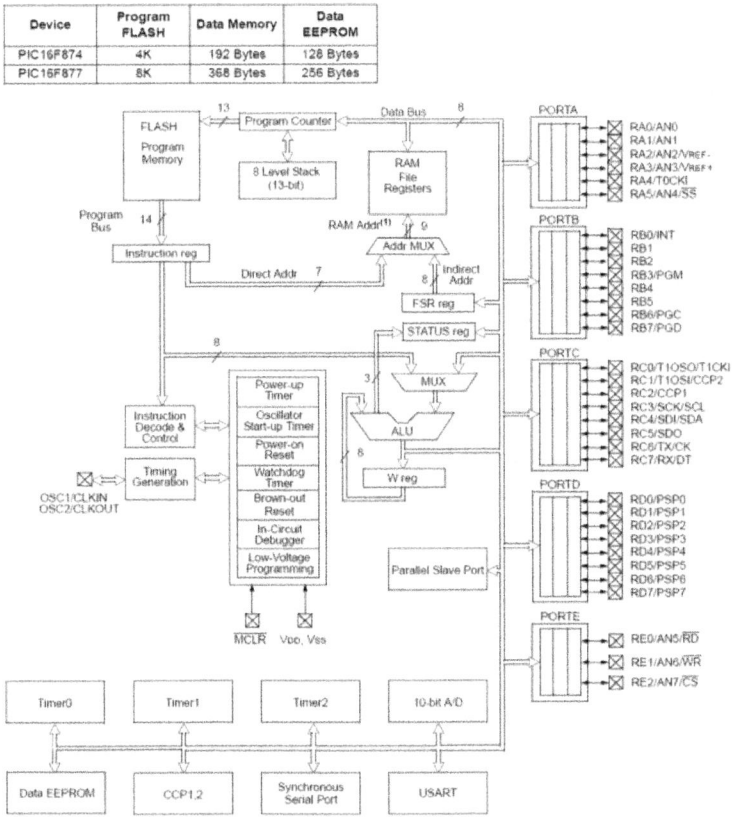

Figure 1: Microcontroller set [7].

Figure 2: EasyPIC-6 cart by MikroElektronika [11].

so learning of new tool interface is minimized. These are the development capabilities of the MPLAB IDE:

- Source-code editing;

- Project management;

- Machine-code generation (from assembly or "C");

- Device simulation;

- Device emulation;

- Device programming.

The comprehensive tool suite allows complete project development without leaving the MPLAB environment [12]. The MPLAB IDE software eases software development as never before in 8-bit microcontroller. MPLAB is a Windows application that contains:

- A full-features editor;

- Three operating modes:

- Editor

- Emulator

- Simulator o A project manager;

- Extensive online help;

MPLAB allows o Editing of source files (ASM and C files);

- One-touch assembly (or compiling) and download to PIC16/17 tools;

- Debugging via:

- Source files

- Absolute listing file

- Program memory o Run-up to four emulators on the same PC;

- Run or single-step;

- Program memory

- Source file

Absolute listing The microchip simulator, MPLAB-SIM, operates under the same platform as the PICMASTER emulator, so the user need only learn a single tool set that functions equally in both the simulator and the full-features emulator [13].

MPLAB-SIM Simulator Software

The software simulator is a no-cost tool for evaluating Microchip's products and designs. Its use greatly helps debug software, particularly algorithms. Depending on a project's design complexity, a time/cost benefit comparing simulator with emulator should be looked into. Projects with multiple development engineers can keep costs down by using both simulator and emulator, allowing speedy debugging of tough problems. MPLAB-SIM Simulator simulates PICmicro series microcontrollers at instruction level. With any given instruction, the user may examine or modify any of the data or provide external stimulus to any of the pins. The input/output radix can be set by the user, the execution performed as either single step, execute until break, or trace. MPLAB-SIM supports symbolic debugging via MPLAB-C and MPASM. The software simulator's low-cost flexibility in developing and debugging code outside laboratory environment makes it excellent multi-project development tool [14,15].

PIC ranges very broadly, from tiny 6-pin 8-bit devices with just 16 bytes of data memory performing only basic digital I/O, to 100-pin 32-bit devices with 512 kilobytes of memory and many integrated peripherals for communications, data acquisition, and control. Newcomers may be confused by an aspect of PIC programming: the lowend devices have entirely separate addresses and data buses for data and program instructions. 8-bit or 16-bit refers to the amount of data that can be processed at once, i.e., the width of the data memory (in microchip terminology, "registers") and the ALU (Arithmetic and Logic Unit). Low-end PICs, operating 8-bit data at any one time, have three architectural families [16,17].

Baseline (12-Bit Instructions)

These PICs are based on the original PIC architecture, going back to the 1970's and General Instrument's "Peripheral Interface Controller". They are rather limited, but within their limits (such as no interrupts) are simple to work with (particularly in modern assemblers such as 6-pin 10F series, 8-pin 12F509, and 14-pin 16F506).

Midrange (14-Bit Instructions)

An extension of the baseline architecture, it supports interrupts, has more memory and on-chip timers and peripherals, includes PWM (pulse width modulation) for motor control, supports serial, I2C, and SPI interfaces, and has LCD controllers. Modern examples include 8-pin 12F629, 20-pin 16F690, and 40-pin 16F887.

High-End (16-Bit Instructions)

Otherwise known as 18F series, this architecture overcomes some limits of the midrange devices. It has more memory (up to 128k program memory and almost 4k data memory) and advanced peripherals (including USB, Ethernet, and CAN or controller area network) connectivity. The 18F architecture supports C programming and is, among 8-bit PIC families, the only one with C compiler. Examples include 18-pin 18F1220, 28-pin 18F2455, and 80-pin 18F8520. Maybe a little confusing is that PIC18F series has 16-bit program instructions operating on 8-bits of data at a time, and is considered an 8-bit chip [12,18]. BASIC programming language is known to users as the easiest and is the most used. The reputation is increasingly transferred onto microcontrollers. PIC BASIC enables quicker and relatively easier program writing for PIC microcontrollers, as compared with Microchip's assembly language MPASM. During program writing, the programmer encounters the same problems always: serial sending of messages, writing of variable on LCD display, generating of six PWM signals, etc. [16].

Facilitating programming are PIC BASIC's built-in commands, which are intended to solve problems typically found in praxis. Where execution speed and program size are concerned, MPASM is less advantaged than PIC BASIC (therefore giving rise to the possibility of combining PIC BASIC and assembler). The part of the program where the same commands are executed many times or the execution time is critical is usually written in assembler. Modern microcontrollers such as PIC execute the instructions in a single cycle lasting 4 tacts of the oscillator. If the microcontroller oscillator is 4 MHz (one tact lasts 250 nS), then one assembler instruction requires 250 nS $\times$ 4 = 1 uS for the execution. Each BASIC command is actually a sequence of assembler instructions; the exact time necessary for execution of a BASIC command is simply the sum of the times necessary for execution of assembler instructions within one BASIC command [17,19].

HARDWARE DESIGN OF THE PIC TRAINER MODEL

Figure 3 is a laboratory model of the trainer design hardware. The model has three main parts: board for applied experiments, PIC microcontroller simulator, and interfacing board with PC computer. There is also a built-in power supply.

Applied Experiments Board

The trainer can execute many experiments: electronic, control logic circuit, power system, etc. The main circuit connecting the board in standalone form will hereby be described. Sensor: the trainer has two types of sensors. One is an approximate sensor (model TURCK Bi14-cp23 APcx sn: 15 mm) detecting front iron paces with 15 mm accuracy and is the approximate switch. Another is photoelectric sensors (serial numbers BR56-DDT-P and BEN9M-TFR). Whereas the former detects interrupts within 5 m, the latter detects reflection two ways: normal closed or normal open. Relays: two types are used, i.e., two poles and three poles, with 24VDC and 24VDC/5A supplying the coils. Four 7-segment models are supplied by 5VDC/2A. Keypads: a matrix of LEDs, with matrix form through logic gates to run instructions of the (i, j) form. The matrix instructions are PIC-programmed and then entered as four rows and four columns. Conveyer belt: supplied by 5VDC/2A, displaying experiment outputs based on sensors or any other processes. DC Motor (model GMN-3M027A/DC24V): its circuit drive executes start/stop, opposite direction, and emergency shutdown instructions; each event is indicated by LEDs with shift rotating. See Figure 4 for the experiment board.

PIC Simulator Board and Interfacing

Five input ports (A→E) and two output ports (T0, T1). The ports transfer instructions and receive sensed signals from the experiment board. Every input/output signal on this board is LED-indicated for ON and OFF. The power supplies are 5VDC, 12VDC, and 24VDC. The PIC simulator is supplied first by the 5VDC and then by the USB cable through a PC. The experiment software is installed on-board via USB through the PIC simulator; another interface is RS232. Figure 5 shows the interfacing board.

SOFTWARE DESIGN OF THE PIC TRAINER

This design uses BASIC language to implement the trainer's experiments. After the program is written in mikroBasic, it is compiled to the PIC. The PC runs the BASIC

Figure 3: The PIC-microcontroller-based conventional electronic control trainer.

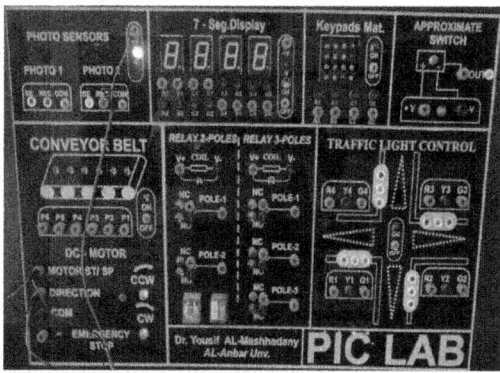

Figure 4: The experiment board.

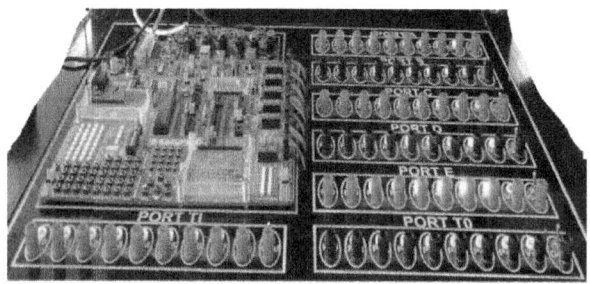

Figure 5: The interfacing board.

compiler program, which translates the original BASIC code into the language of 0s and 1s understood by the microcontroller. Figure 6 shows the translation of a BASIC program into an executive HEX code. The program, written in PIC BASIC and registered as Program.bas file, is converted into assembler code (Program.asm), which is further translated into executive HEX code written to the microcontroller memory by a programmer (a device transferring HEX files from the PC to the microcontroller's memory). Each experiment has two procedures: one to write the PIC programming code by software, another to implement the hardware connection.

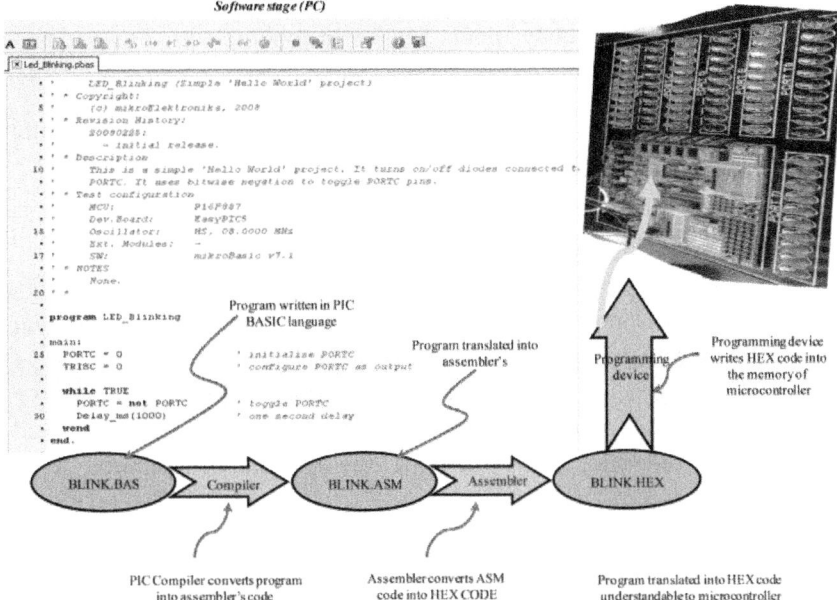

Figure 6: Details of the PIC trainer software.

CASE STUDY OF THE TRAINER'S USE

The trainer was designed to implement experiments of various electrical engineering fields. It is capable of highlevel research projects and can be used in undergraduate laboratories. Two case studies, for power and electronic, are presented: DC motor controller and intelligent traffic light.

Controller for DC Motor: in the experiment, three main operations (start/ stop, control of clockwise and anticlockwise directions, and emergency shutdown) are applied to a 24VDC motor (see Figure 7 for the drive circuit of the three operations). A 24VDC one-pole relay was used. The circuit could be manually controlled and could also use a PIC microcontroller to operate relay coil for executing a suitable instruction to the DC motor.

- The experiment procedure is:

- Connect the board section (see Figure 8) to the three external power supplies on-board the microcontroller.

- Connect the other details to the microcontroller's output pinnae.

- Write a program for the three experiment parts and set the program on a PIC chip (any serial, e.g., 16F- 667, 16F84A, etc.) and arrange for port A of the microcontroller to be the output.

- Feed the emergency inputs sensed by photoelectric sensor; manual feeding is possible as needed.

Run the circuit by power ON of the voltage source and examine the instruction to the DC motor.

The output must be present as the motor shift rotating and the corresponding LED lighting up showing the direction of rotation.

Intelligent Traffic Light: Optimal waiting time for traffic lights to change will reduce carbon monoxide emission, also save motorists' time and reduce frustration. Other advantages are no interference between the sensor rays and no redundant signal triggering. Ability to interface with software allows this sensor-based traffic system to easily accept feedback (the software and the hardware can communicate). Table 1 lists the operation sequences.

The experiment procedure is:

Ø   Connect the section of the board (see Figure 9) to output ports A and B of the microcontroller board.

Ø   Write a program for the three parts of the experiment and set the program on the PIC chip (PIC 16F667) and arrange the output to be at ports A and B of the microcontroller.

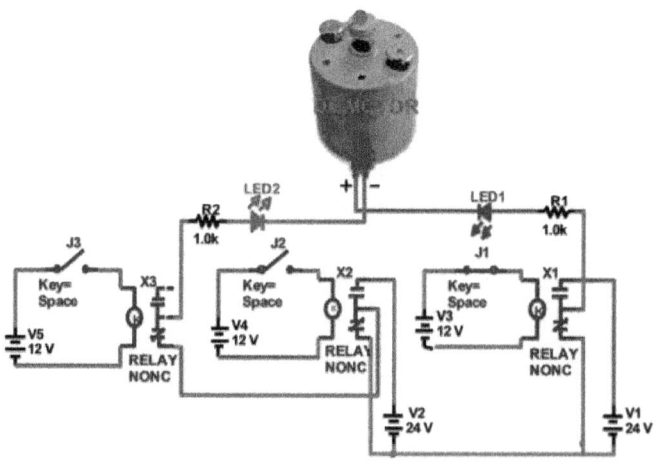

Figure 7: The drive circuit in the PIC controller experiment with the DC motor.

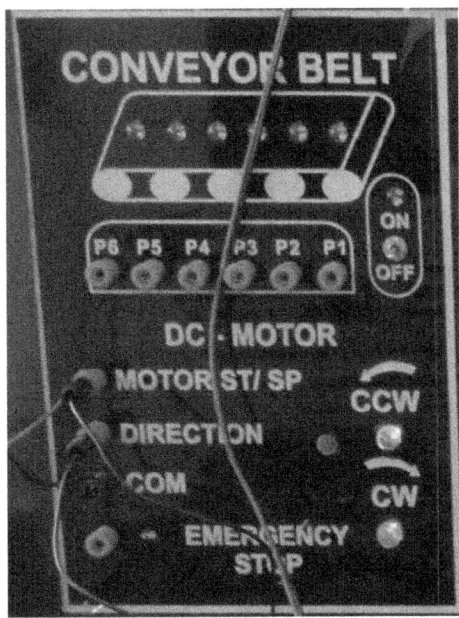

Figure 8: The DC motor section of the experiment.

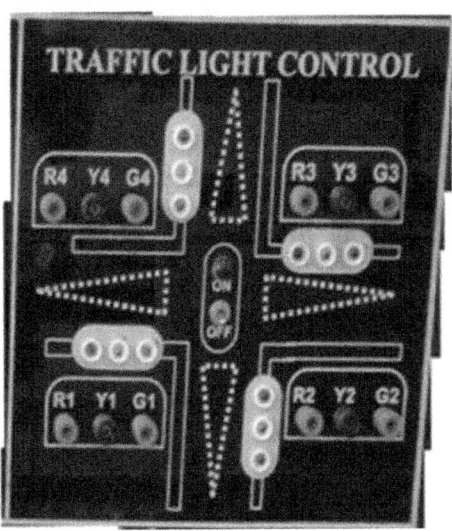

Figure 9: Traffic light experiment schematic.

- Turn ON the power of the main board and then of the microcontroller board.
- Record the lighting time and sequence and compare with the program.

Table 1: Operation sequences of the intelligent traffic light

No.	Time	Traffic 1	Traffic 2	Traffic 3	Traffic 4
State 1	0 to 10 sec	Green 1	Red 2	Red 3	Red 4
	11 to 15 sec	Yellow 1	Red 2	Red 3	Red 4
State 2	16 to 25 sec	Red 1	Green 2	Red 3	Red 4
	31 to 40 sec	Red 1	Yellow 2	Red 3	Red 4
State 3	41 to 45 sec	Red 1	Red 2	Green 3	Red 4
	46 to 50 sec	Red 1	Red 2	Yellow 3	Red 4
State 4	51 to 60 sec	Red 1	Red 2	Red 3	Green 4
	61 to 65 sec	Red 1	Red 2	Red 3	Yellow 4
Return to State 1	66 to 75 sec	Green 1	Red 2	Red 3	Red 4
and so on...	76 to 80 sec	Yellow 1	Red 2	Red 3	Red 4

- Investigate the effect of using photoelectric sensor signal; can it show what happens at a street intersection?

CONCLUSIONS

Results from practical implementation of all the experiments and simulation results from programming the PIC microcontroller board show the trainer to be very useful and necessary to many design plans. Its higher performance, lower cost, higher accuracy, and better speed response are all as compared with many types of classical trainers for electronic and control systems.

Its facilities will shorten the time taken for many design procedures (where applicable), simulations, and experiments; each can also be an individual system. The design enables instant initial results and modification of experiment steps such as setting the initial condition and updating some of the parameters, so the trainer's accuracy and performance are increased. The trainer allows practical simulations of many real systems. Capable of a wide range of experiments, it is very suitable for use in higher education laboratories. New experiments can be included by adding new circuits to the board and rearranging the connections.

ACKNOWLEDGEMENTS

Al Anbar University supported the project through manufacturing laboratory instruments funding. Al Sofa office provided help with consulting notes for the microcontroller's practical application.

REFERENCES

1. N. Barsoum, "Speed Control of the Induction Drive by Temperature and Light Sensors via PIC," Transaction in Controllers and Drives, 2010, pp. 35-59.

2. M. Bates, "Interfacing PIC Microcontrollers Embedded Design by Interactive Simulation," Elsevier, Amsterdam, 2006.

3. F. J. Diaz, F. J. Azcondo, R. Casanueva and C. Branas, "Microcontroller Software Applied to Electronic Ballast Design," 13th European Conference on Power Electronics and Applications, Barcelona, 8-10 September 2009, pp. 1-8.

4. H. W. Huang, "PIC Microcontroller: An Introduction to Software and Hardware Interfacing," 2005. http://www.delmar.com.

5. J. Main, "Measuring Resistance Using Digital I/O Using a Microcontroller for Measuring Resistance without Using an ADC," 2008. http://www. best-microcontroller-projects.com

6. C. Singh and K. Agarwal, "Design of Reactive PIC Microcontroller," Proceedings of International Symposium on Signals, Systems and Electronics, Pilani, 17-20 September 2010, pp. 1-4.

7. H. Rongen, "Introduction to PIC Microcontroller," Forschungszentrum Jülich Zentrallabor für Elektronik, Jülich, 2009.

8. N. Gardner, "An introduction to Programming the Microchip PIC in CCS C," Ccs Inc., Christiansburg, 2002.

9. B. Hossain, N. Hossain, M. Hossen and H. Rahman, "Design and Development of Microcontroller Based Electronic Queue Control Systems," Proceeding of the 2011 IEEE Students' Technology Symposium, Kharagpur, 14- 16 January 2011, pp. 48-52.doi:10.1109/TECHSYM.2011.5783862

10. T. Wilmshurst, "Designing Embedded Systems with PIC Microcontrollers Principles and Applications," Elsevier Ltd., New York, 2007.

11. N. Matic and G. Maneger, "EasyPIC Microcontroller Board User

Manual," 2008.

12. D. Ibrahim, "Advanced PIC Microcontroller Projects in C from USB to RTOS with the PIC18F Series," Elsevier, Amsterdam, 2008.

13. F. H. Fahmy, S. M. Sadek, N. M. Ahamed, M. B. Zahran, and A. El-S. A. Nafeh, "Microcontroller-Based Moving Message Display Powered by Photovoltaic Energy," International Conference on Renewable Energies and Power Quality, Granada, 23-25 March 2010. http://www.icrepq.com/icrepq'10/726-Sadek.pdf

14. Y. Aye, "Design and Construction of LAN Based Car Traffic Control System," World Academy of Science, Engineering and Technology, Vol. 46, 2010, pp. 586-591.

15. N. Matic, "BASIC for PIC Microcontrollers," 2001. http://scalak.elektroda.eu/html/pliki/BasicforPICMicrocontrollers.pdf

16. F. J. Díaz, F. J. Azcondo, R. Casanueva and Ch. Brañas, "Microcontroller Software Applied to Electronic Ballast Design," University of Cantabria, Cantabria, 2010.

17. S. C. Hsiung, "The Use of PIC Microcontrollers in Multiple DC Motors Control Applications," Journal of Industrial Technology, Vol. 23, No. 3, 2007, pp. 2-3.

18. S. Huseinbegovic and O. Tanovic, "Development of a Distributed Elevator Control System Based on the Microcontroller PIC 18F458," 2010 IEEE Region 8 International Conference on Computational Technologies in Electrical and Electronics Engineering, Irkutsk, 11-15 July 2010, pp. 858-863.

19. L. D. Jasio, et al., "PIC Microcontrollers," Elsevier Inc., New York, 2008.

Chapter 9

PROFESSIONAL STRESS IN JOURNALISM: A STUDY ON ELECTRONIC MEDIA JOURNALISTS OF BANGLADESH

Kazi Nazmul Huda[1], Abul Kalam Azad[2]

[1]Department of Business Administration, Southern University Bangladesh, Chittagong, Bangladesh

[2]Department of Communication & Journalism, University of Chittagong, Chittagong, Bangladesh

ABSTRACT

To meet the objectives of professional assignments, journalists often suffer from high level professional stress. The key objective of this study is to identify the major determinants of professional stress of journalists in electronic media. The study is, to a large extent, based on quantitative data collected through personal interview of 55 journalists of different private television channels of Bangladesh. The study reveals that inadequate support from management is the most heated cause of professional stress. However, the result also concludes "harassment" at workplace as an insignificant determinant. Other major determinants of stress are "unclear objective", "insecure job climate", "excessive time pressure", "friction", "long working hours" and "life threat" found in the study. However, though the study is very new to Bangladesh perspective, it is expected that it may help to develop professional stress management policies in this sector.

INTRODUCTION

Job is an important part for many people's life and professional stress related to job is inevitable. There is a firm relation between profession-related stress and performance and there is evidence to support that the stress affects organizational and individual productivity (Bradley & Sutherland, 1994). It is a contemporary issue both for the employees and employers and has become an unavoidable factor in the high performance corporate culture. There are many kinds of stresses in human life; but professional stress or the one that is related to job is unique in nature and may become "the silent killer" if not managed properly (Tarkovsky, 2007). Professional stress causes different kinds of physiological, psychological, and behavioral symptoms (Table 1) that may lead to collapse of human machine. Organizations and employees must deal with the issues of professional stress and should be smart in tackling stress to stay productive. Employees should be knowledgeable regarding the causes to meet professional stress on time. Professional stress occurs when the employees find disparity between job and work environment especially when they need to work under different categories of pressures and anxieties (Nawe, 1995 ; Maslach, 2003). Continuing a stressful job may lead to sudden burnout situation at workplace, and could develop negative self-image and other symptoms of stress (Caputo, 1991).

The press serves the role of delivering new information quickly, accurately, and consistently (Sang-young & Cho, 2014) and the electronic media in Bangladesh has grown rapidly and currently, playing a significant role in contributing to the entertainment industry. Since 1997, the industry witnessed a significant qualitative and quantitative change when this sector was privatized. After the independence of Bangladesh in 1971, there was only TV channel in the country namely Bangladesh Television (BTV) run by the government with limited number of programs and news. However, since 1997 the industry has achieved significant expansion with 41 private television channels. News is a common product of all TV channels and due to more competition caused by the increasing number of television channels (Azad & Hussain, 2015) , enormous stress on the news reporters is given.

Journalism is one of the most stressful careers that have to deal with deadlines, busy work environments, tight schedules, extensive travelling, fulfilling the demands of the editors, and the fear of being killed and laid off (Shmoop, 2014). The profession is been rated in the top ten stressful jobs amongst all the professions in the world and faces a shrinking job market (Shapley, 2013). A large number of news reporters in Bangladesh had complained that they could not even report fair news by the threats and pressure of employers and effected parties (Azad & Hussain, 2014). Journalists are often subject to dangerous and life threatening situations and encounter considerable amount of psychological pressure besides other stress, i.e. work overload and job insecurity. Existing studies hardly succeeded in finding out finding out the stress depicts, and rarely generated many ideas to mitigate professional stress (Teasdale, 2006). Our study intends to close the gap.

When professional stress takes a massive shape, it affects the performance of the organization, turns down the quality of individual productivity, deteriorates physical condition, weakens social relations, and ruins family life. Stress at work is inevitable in today's high performing and competitive corporate culture. However, active interventions may minimize the level of stress to bring a balance between work and life (Baker, 1985). It is the responsibility of the employers and employees to identify the causes of professional stress and should measure its impact on performance to maintain a healthy and productive work life and to avoid the negative consequences. The study tires to address the specific causes of stress of electronic media journalists of Bangladeshi private television channels and endeavors to reveal their opinions about the determinants of professional stress.

OBJECTIVES OF THE STUDY

1) To identify and rank the determinants of professional stress of electronic media journalists.

2) To recommend specific interventions to minimize professional stress of them.

Table 1: Types of symptoms of professional stress

Sl. No.	Physiological	Mental/ Psychological	Behavioral
1	Headaches	Anxiety	Irritable/aggressive
2	Migraines	Tension	Withdrawn
3	Changing sleep patterns	Irritability	Social isolation
4	Stomach disorders	Low self-esteem	Consumption of alcohol, tobacco, tea/coffee, drugs, self-medication
5	Raised blood pressure	Forgetfulness	Non co-operative
6	Muscle spasms	Feeling powerless	Accident-prone
7	Back/shoulder/neck pain		Less sociable
8	Sense of feeling unwell		Negligence in personal appearance
9	Unwillingness to work		

Source: Newstrom and Davis (2007) and Authors.

LITERATURE REVIEW

Professional stress could be a chronic life-threatening syndrome raised from the work conditions that may negatively influence employees' productivity and personal well-being. According to World Health Organization (1948) stress are states of comprehensive physical, mental, and social illness in a person. Though, no profession is stress-free; and has some degree of strain, and anxiety that could result in productivity and satisfaction at work or may lead to negative results like mental and physical illness if the stress is excessive

(Teasdale, 2006). Work-related stress is considered as the foremost cause of a wide range health problems (Kivimaki et al., 2006), and is strongly connected to staff turnover, absenteeism, poor morality and declining productivity (Noblet & Lamontagne, 2006). Increasing level of workplace stress can lead to serious legal allegation against any employers like compensation claims, disciplinary issues, and workplace violence Brady (cited in Babcock, 2009) . According to Canadian Underwriter (2004) , factors that causes professional stress include conflict/friction among the co-worker (University of Cambridge, 2014 ; Friedman et al., 2000), bulling by the supervisors, job insecurity, and the absence of freedom in decision-making (University of Cambridge, 2014), and some personal issues i.e. family pressure, financial constrains abuse etc. It could be also generated out of fear of uncertainty, unrealistic deadlines and interpersonal conflict (Babcock, 2009). Professional stress mostly occurs by long working hours that may cause cardiovascular attack (Uehata, 1991). It also interferes family life and psychological distress (Major et al., 2002). In many times, bullying and organizational incivility are one of the most upsetting issues for the employees (Gholipour et al., 2011), that might be caused by allegation, rudeness, frightening, malevolence, insult which directs to aggravation, threat, disrespect, and deterioration of self-confidence (Lee, 2000). However, the more common psychological job stress is anxiety, and depression that negatively impacts work environment (Teasdale, 2006). As stated in the U.S Department of Health, Professional stress is harmful physical and emotional responses when the job description of a job does not match the capabilities, resources, or needs of the employees. The fight to stabilize the work and family life is purely one of the many stressors that an employee faces at work (Tyler, 2006). In many times professional stress are created by pressure of ethical conduct (Ulrich et al., 2007 ; Glicken, 2013). To combat the stress the managers must audit and acknowledge the presence of stress at workplace should undertake stress management interventions to reduce the levels of stress of their employees (Sidle, 2008). According to Jones et al. (2003), it is found that stress management interventions improve physical and mental health, reduces costs of the employers, and facilitates the reintegration of effected employees into workplace and it is an integral component of health promotion program of an organization (Kobayashi, 1997). There are three broad categories of stress management interventions exposed by Ivanevich et al. (1990) , i.e. reducing the current stressors, identifying the employees under stress and aiding the employees to adjust with the situation that causing stress. DeFrank and Cooper, (1987) recommended, stress management as individual interventions like relaxation and autonomy at work. Stress management techniques could be divided into two types: environmental management (Murphy, 1999) which is an endeavor to organize work environments to reduce the cause of stress;

and the approach that aids the employees to deal efficiently with various types of stressful condition (Hardy & Barkham, 1999). Physical and mental health is a potential source of work related injury or disease and both the employer and employee have the lawful right, responsibility and accountability to help each other to identify, acknowledge and mitigate stress and stressors to ensure sustainable productivity and workplace harmony.

METHODOLOGY, SCOPE AND LIMITATION OF THE STUDY

The methodology of this study was designed to address of objectives of the research. The study is descriptive in nature, followed inductive research approach and a survey based research strategy. The methodological choice of the research is both qualitative and quantitative. The survey intervenes to collect required information and data through structured questionnaire. An extensive literature review was conducted to identify the most common determinants of professional stress (Table 2) and the questionnaire was developed based on the identified variables. Total 55 journalists of fourteen private TV channels were interviewed with a view to making the study informative and representative and a close-end questionnaire survey was conducted too. The questionnaire contains 15 questions and a 5-point likert rating scale (5 Strongly agree…1 Strongly disagree) were used to capture the opinion of the respondents about important determinants of professional stress. The respondents were mostly male news reporters and camerapersons who have permanently employed in this sector for last five years. The survey was carried out during May to June 2015.

Table 2: Important determinants of professional stress

	Determinant of Professional Stress	Code	Source
1	Unclear objective	Objective	Semmer (2007)
2	Excessive time pressure	Time pressure	Semmer (2007)
3	Unachievable deadline	Deadline	Turnage & Spielberger (1991); University of Cambridge (2014)
4	Long work hour	Work hour	Sethi et al. (2004); Uehata (1991); University of Cambridge (2014)
5	Work overload	Overload	Sethi et al. (2004); Semmer (2007); University of Cambridge (2014)
6	Fast work	Fast	Sethi et al. (2004); Major et al. (2002)
7	Less freedom	Freedom	University of Cambridge (2014); Canadian underwriter (2004)
8	Inadequate support	Support	Semmer (2007)
9	Workplace harassment	Harassment	Semmer (2007)
10	Friction	Friction	University of Cambridge (2014); Friedman et al. (2000)
11	Pressure for un-ethical conducts	Ethics	Ulrich et al. (2007); Glicken (2013)
12	Insufficient compensation package	Compensation	Canadian underwriter (2004)
13	Family pressure	Family	Major et al. (2002); Canadian underwriter (2004)
14	Possibility of life threat	life threat	Zakaria & Azad (2009)
15	Insecure job climate	Job security	Semmer (2007); University of Cambridge (2014)

Source: Literature Review.

Data analysis of the study used intensive statistical tool such as multivariate analysis technique, specifically Exploratory Factor Analysis (EFA) with the support of SPSS. Important determinants of professional stress have been identified by considering the loading value of each variable (Table 2). Higher loading value is considered as high importance of professional stress and vice versa. The respondents were asked about their feelings of stress in terms of the variables, i.e. are they clear about their job objectives? Do they feel time pressure? Are they overloaded with responsibilities? Is there a friction among the peers? What about their feelings of job security. Are they under life threat?, and so on. The researchers have their level best to cover up many issues regarding the professional stress of journalists. However, the study has some limitations too. Major limitation of the study is the scope of the research, as the area does not cover all the television channels of Bangladesh due to financial constrain of the researcher. The sample size was limited to 55 as the research was supposed to be completed in a limited period. The demographic data of the respondents, i.e. age, gender, and others are not considered in the study. The researcher was bias to theories of stress management only to develop the model of the research and interpretation & analysis of the data were organized in accordance to that. The possibility of respondent's responses being biased cannot be ruled out too.

FINDINGS

The key objective of the study was to identify the important determinants of professional stress among electronic media journalists of Bangladeshi private television channels. In order to fulfill the research objective, and at the very outset, factor analysis was conducted, as it is an effective statistical tool used to describe variability among observed, correlated variables in terms of a potentially lower number of unobserved variables. Principal Component Analysis was conducted on fifteen (15) variables (Table 3) where the KMO value of study was found to be 0.477 and the TVE value is 67.25% at the first stage of EFA. The result shows that, these variables in combined can explain more than 67% of the total variation of professional stress. However, the variable namely "workplace harassment" was found to be cross-loaded (loading in two dimensions) and that might have caused the lower KMO value. That is why the variable "harassment" is dropped from the study. Workplace harassment may not be an important determinant among the reporters and camerapersons as most of the time they work outdoor and have less formal interaction and indoor responsibilities. However, crime and corruption reporters face somewhat harassment outdoor by the parties affected by their news reports.

After dropping harassment from the study, KMO value has increased from 0.47 to 0.512 and which is significant too. This means the problem initially was might be due to less important variable, harassment. Principal

Table 3: Factor summary

Factor	Variables	Loading Value	Eigenvalue	Percentage of Variation Explained
1	Insecure job CLIMATE Fast work Less freedom	0.838 0.692 0.601	2.530	18.072
2	Friction Family pressure Insufficient compensation package	0.800 0.691 0.618	2.015	14.395
3	Excessive time pressure Unachievable deadline	0.834 0.615	1.529	10.925
4	Long working hours Work overload Pressure for un-ethical conducts	0.742 0.661 0.606	1.468	10.485
5	Unclear objective Possibility of life threat	0.849 0.710	1.102	7.847
6	Inadequate support	0.888	1.026	7.329

Source: Compiled from SPSS Version 20.

Component Analysis have clustered remaining 14 variable under six broad factors and these factors covered around 69% of the total variance of the professional stress. Factor loading of the variables determining the degree of significance of each factor and the Eigenvalue and percentage of variation explained by the factor are shown in Table 3. The given result provides statistical evidence to support the newly identified six factors of professional stress factors as coded F1, F2, F3, F4, F5 and F6 (Table 3). This shows a firm influence of these variables to cause professional stress as these variables covered 69.081 variances (Appendix 1). Under factor 1, insecure job climate, fast work and less freedom determinants are clustered with the eginenvalue of 2.530 and the percentage of variation explained 18.072 that show significant influence in causing professional stress. The factor may cause both mental and physical stress in professional life. Factor 2 contains three variables friction, Family Pressure, and insufficient compensation package with the eigenvalue 2.015 and the percentage of variation explained 14.395. This factor mostly creates mental and behavioral stress. Factor 3 contains the variables, such as excessive time pressure, and unachievable deadline. The eigenvalue of this dimension is found to be 1.529 and the percentage of variation explained at 10.925. This factor is mostly responsible in causing physical and mental stress in the profession. Factor 4 encloses three variables termed as long work hour, work overload, and Pressure for un-ethical conducts. The eginvalue of this factor is 1.468 and the percentage of variation explained 10.485. Presence of this factor at profession may cause physical, mental, and behavioral stress.

Variables enfold in Factor 5 namely unclear objective and life threat with the eginvalue 1.102 and the percentage of variation explained 7.847. The factor might cause physical and mental stress in the profession. Factor 6 contains only one determinants inadequate support having the eginvalue 1.026 and the percentage of variation explained 7.329. This factor may lead to causing mental and behavioral stress in the profession. Though loading table shows that, 14 variables are grouped under six dimensions/factor (Table 4 and Table 5), but the plot diagram (Figure 1) argues that they are close variables and somewhat uni-dimensional as we see the circles and their close proximity in the diagram. This means maximum determinants of professional stress are uni-dimensional and interrelated to each other in causing stress.

DISCUSSION

The study shows a clear picture of the presences of professional stress in electronic media journalism and identified the influencing determinants that causing stress in their professional life. The main findings of the study are:

1) The six factors that are found out of 14 variables.

2) Loading values of the variables are greater than 0.50 and found positive in relation to stress.

3) Some having highly significant relationship, as their loading value is more than 0.80. Highest loading vale amongst 14 purified variables is "inadequate support" (loading is .888). It means the journalist needs support from many dimensions like from the management, peer group, subordinates to cover and broadcast news and a clear indication to the necessity of teamwork in this profession is quite obvious. The journalist may need external

Table 4: Total variance explained

Component	Initial Eigenvalues			Rotation Sums of Squared Loadings		
	Total	% of Variance	Cumulative %	Total	% of Variance	Cumulative %
1	2.530	18.072	18.072	1.875	13.394	13.394
2	2.015	14.395	32.468	1.734	12.387	25.781
3	1.529	10.925	43.393	1.708	12.196	37.977
4	1.468	10.485	53.878	1.590	11.357	49.335
5	1.102	7.874	61.752	1.463	10.451	59.785
6	1.026	7.329	69.081	1.301	9.296	69.081
7	0.925	6.607	75.687			
8	0.829	5.925	81.612			
9	0.677	4.834	86.446			
10	0.550	3.928	90.374			
11	0.408	2.913	93.287			
12	0.382	2.728	96.015			
13	0.304	2.175	98.190			
14	0.253	1.810	100.000			

Extraction Method: Principal Component Analysis.

Table 5: Rotated component matrix[a]

	Component					
	1	2	3	4	5	6
Unclear objective					0.849	
Excessive time pressure			0.834			
Unachievable deadline			0.615			
Long work hour				0.742		
Work overload				0.661		
Fast work	0.692					
Less freedom	0.601					
Inadequate Support						0.888
Friction		0.800				
Pressure for un-ethical conducts				0.606		
Insufficient compensation package		0.618				
Family pressure		0.691				
Possibility of life threat					0.710	
Insecure job climate	0.838					

Extraction Method: Principal Component Analysis. Rotation Method: Varimax with Kaiser Normalization. [a]Rotation converged in 8 iterations.

supports from the governmental authorities and stakeholders of media like politicians, civil societies, and human right activists to perform their professional responsibility fairly and freely.

The second most important determinants of stress is "unclear job objectives" with the loading value of 0.849. It means significant stress can be created if the job objective is not making clear to the journalists. Objective in journalism can be of two types, one is professional objective, and another is task related objective that is time bound. The first situation of unclear objective may occur if they feel or find a gap between the dignity of the profession and the reality of work life. Because in real life, it is very tough to broadcast true news with the possibility of life threat (0.710), less freedom to work (0.601), pressure for unethical conduct (0.606). Zakaria & Azad (2009) also found the same in their study and stated that, this profession is full of stress with life threat, unethical pressure by the owners as most of them are politically connected and therefore, put pressure on the journalist to perform true responsibility. The second situation of unclear objective may occur mostly for the presences of factor 2 where the fourth highest loading value excessive time pressure is present with 0.834 because the work objective means, task to be finished in a given time.

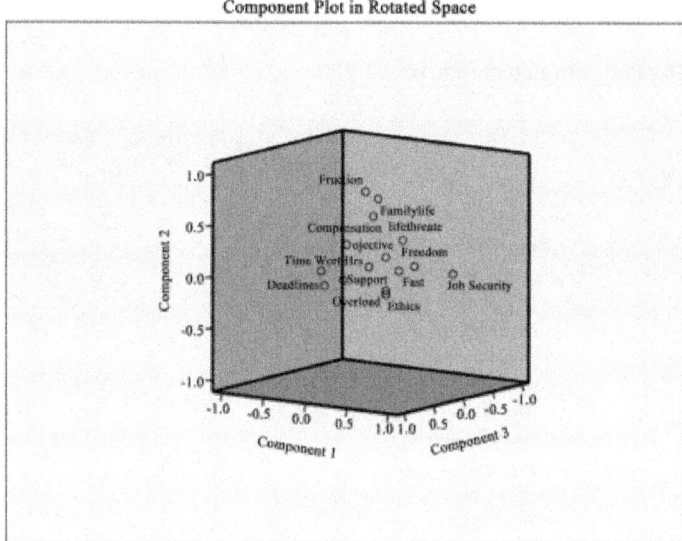

Figure 1: Plot diagram.

Third highest variable is "insecure job climate" with the loading value of 0.838. The journalist could be feeling job insecurity as the privet job culture in Bangladesh is not yet matured with the concept of employee retention strategies. Zakaria & Azad (2009) also found the same findings and stated that this profession is full of uncertainty, with no job security, and no human resource policy and rules. Friction is common in this profession where greater degree of interdependency is required to accomplish exclusive assignments. Every news is a unique project work and there is a presence of aggressive competition among the reporters for the priority of broadcasting may create a competitive environment that may lead to friction among the peers and superiors. So, we see a higher loading value of the variable "friction" (0.800) in the study. Bangladesh frequently faced politically turmoil situations and the TV reporters had to run after breaking news and needed to work restless for long hours. That is why "long working hour" with the loading of 0.742 is another important determinant of professional stress to them. From the study, it is revealed that, stress in electronic media journalism is caused by multiple determinants and they are interrelated, too.

RECOMMENDATIONS

Professional stress is not determined by only job related stress. There is an existence of external stressors like, family pressure, life threat, and unethical environment. Therefore, a holistic approach could be taken by the top management of the TV channels to design effective interventions in identifying and reducing the important determinants of professional stress. Tailored-made stress management programs like stress reporting and auditing, job redesign, grievances management program, counseling, meditation and psychotherapy could be under take to bring stability in this noble profession and support the journalist to cope with the stressors. Sidle (2008) and Bradley & Sutherland (1994) suggested the same in their study. Stabilizing functions should act to sustain the process continuity through periodic stress audit and feedback.

At the beginning, awareness program on professional stress could be organized to aid the management and the journalists to take control over work and life. Encourage peer support culture among the journalists through promoting team culture in this profession. Reporters could be empowered with greater flexibility in decision-making and due assistance to attain their objective without stress. Effective teamwork will allow the journalists to share responsibilities and value each other priority and develop a caring and supportive environment to accomplish stressful assignments. Bradley & Sutherland (1994) also suggested conducting training on teambuilding and relaxation. Some specific stress management interventions like relaxation program to be arranged for the journalists. Most of the research on stress management discussed the success of relaxation program in combating stress. Relaxation program may include yearly outdoor family events could be organized to keep them charged and stress free. Employee wellbeing program, i.e. recreations facility, family events, family friendly work environment, recognition, and reward for their wellness.Teasdale (2006) also mentioned to bring a balance in professional life. Work life balance issue is to be considered as a key issue and employee concierge service could be introduced to help busy employees to make their family life easy. This employee assistance service may include household services, and onsite medical care, utility bill payment, repair services, banking services, legal advice services etc to reduce off-the-job concerns, increase on-the-job focus, and company loyalty, to balance work and personal lives and enjoy hassle free and peaceful life. Some of the government organizations Bangladesh, like military services, organizations under petroleum corporations, port authorities, are provided with these types of services.

Press Club, the association of the journalists in Bangladesh, may take some measures to help the journalist by taking programs like counseling, advocacy program, policy formulation to reduce stress and mainly a unified service rules for the journalists to promote better human resource management practices in this profession. Government should undertake protection programs for the journalist by providing security by law enforcing agencies. The journalists killed in the line of duty should be provided with justice and their family members should be taken under governmental rehabilitation program.

CONCLUSIONS

Electronic media journalists are playing a pivotal role in promoting democracy, human rights, good governance, and other social issues. So, their stress in the professional life should be assessed with due diligence. This study finds the presence of professional stress among the Bangladeshi electronic media journalists with a significant value. However, the professional stress of TV journalists are largely dependent on less support, unclear objective, insecure job climate, time pressure, friction, long-hour work and a new discovery that is life threat. Harassment does not come as an important determinant in this profession but external bodies usually harass them. Adequate support, job security, positive interpersonal relationship among the co-employees, better salary, and fringe benefits could bring higher efficiency in this profession. However, the journalists should take some personal initiatives to be knowledgeable about the ways of handling professional stress and develop a self-help approach to reduce the degree of stress to remain productivity as well. Journalism has become a critical occupation day by day. In search for news stories, they need to see, write, and witness many unpleasant incidents that could be stressful and may end with life threatening consequence of their life. A well thought-through, proactive approach of professional stress management could be organized by all the stakeholders of journalism to combat stress and to make this challenging profession pressure-free, productive, and enjoyable.

Professional stress is a multi-dimensional situation of a person's life that could be caused by off and on the job stressors. In managing people with productivity and job satisfaction, professional stress management is a prime issue of consensus and consciousness for the new generation managers, employees, and the policy makers at large. The concept of professional stress and its major determinants could be an important issue of study to bring efficiency in work and balance in social life. The research could be helpful to identify the symptoms of stress, outcome, and impact of stress on performance of the journalists in the future.

REFERENCES

1. Azad, A. K., & Hussain, S. (2015). IBS Journal, Vol. 22. Bangladesh: Institute of Bangladesh Studies.

2. Baker, D. B. (1985). The Study of Stress at Work. Annual Review of Public Health, 6, 367-381. http://dx.doi.org/10.1146/annurev.

pu.06.050185.002055

3.   Bradley, J., & Sutherland, V. (1994). Stress Management in the Workplace Taking Employees' Views into Account. Employee Counseling Today, 6, 4-9.http://dx.doi.org/10.1108/13665629410060443

4.   Brady, J. (Cited from Babcock, P.) (2009). Workplace Stress? Deal with It. HR Magazine, May 2009, 67.

5.   Canadian Underwriter (2004). Stress from Workplace Conflict.www.canadianunderwriter.ca

6.   Caputo, J. (1991). Stress and Burnout in Library Service. Phoenix, AZ: Oryx Press.

7.   DeFrank, R., & Cooper, C. (1987). Worksite Stress Management Interventions: Their Effectiveness and Conceptualization. Journal of Managerial Psychology, 2, 4-10.http://dx.doi.org/10.1108/eb043385

8.   Friedman, R. A., Tidd, S. T., Currall, S. C., & Tsai, J. C. (2000). What Goes around Comes around: The Impact of Personal Conflict Style on Work Conflict and Stress. International Journal of Conflict Management, 11, 32-55.http://dx.doi.org/10.1108/eb022834

9.   Gholipour, A., Sanjari, S., Bod, M., & Kozekanan, S. (2011). Organizational Bullying and Women Stress in Workplace. International Journal of Business and Management, 6, 234. http://dx.doi.org/10.5539/ijbm.v6n6p234

10.   Glicken, M. D. (2013). Treating Worker Dissatisfaction in a Time of Economic Change. Elsevier

11.   Hardy, G., & Barkham, M. (1999). Psychotherapeutic Interventions for Work Stress. In J. Firth-Cozens, & R. L. Payne (Eds.), Stress in Health Professionals: Psychological and Organizational Causes and Interventions (pp. 247-259). Chichester: John Wiley & Sons.

12.   Ivanevich, J., Matteson, M., Freedman, S., & Philips, J. (1990). Worksite Stress Management Interventions. American Psychologist, 45, 252-261. http://dx.doi.org/10.1037/0003-066X.45.2.252

13.   Kivimaki, M., Virtanen, M., & Elovainio, M. (2006). Work Stress in the Etiology of Coronary Heart Disease—A Meta-Analysis. Scandinavian Journal of Work, Environment & Health, 32, 431-442. http://dx.doi.org/10.5271/sjweh.1049

14.   Kobayashi (1997). Japanese Perspective of Future Work Life. Scandinavian Journal of Work, Environment & Health, 23, 66-72.

15.   Lee, D. (2000). An Analysis of Workplace Bullying in the UK. Personnel Review, 20, 593-610. http://dx.doi.org/10.1108/00483480010296410

16. Major, V. S., Klein, K. J., & Ehrhart, M. G. (2002). Work Time, Work Interference with Family, and Psychological Distress. Journal of Applied Psychology, 87, 427-436.http://dx.doi.org/10.1037/0021-9010.87.3.427

17. Maslach, C. (2003). Job Burnout: New Directions in Research and Intervention. Current Directions in Psychological Science, 12, 189-192. http://dx.doi.org/10.1111/1467-8721.01258

18. Murphy, L. (1999). Organizational Interventions to Reduce Stress in Health Care Professionals. In J. Firth-Cozens, & R. L. Payne (Eds.), Stress in Health Professionals: Psychological and Organizational Causes and Interventions (pp. 149-162). Chichester: John Wiley & Sons.

19. Nawe, J. (1995). Work-Related Stress among the Library and Information Workforce. Library Review, 44, 30-37. http://dx.doi. org/10.1108/00242539510093674

20. Newstrom, J., & Davis, K. (2007). Organizational Behavior: Human Behavior at Work. New Delhi: McGraw-Hill.

21. Noblet, A., & Lamontagne, A. D. (2006). The Role of Workplace Health Promotion in Addressing Job Stress. Health Promotion International, 21, 346-353.http://dx.doi.org/10.1093/heapro/dal029

22. Sang-young, P., & Cho, S. (2014). Effects of Journalists' Job Stress Factors on Physical Conditions. Advanced Science and Technology Letters, 72, 11-15

23. Semmer, N. K. (2007). Recognition and Respect (or Lack Thereof) as Predictors of Occupational Health and Well-Being. Paper Presentation at World Health Organization, Geneva: WHO.

24. Sethi, V., King, R. C., & Quick, J. C. (2004). What Causes Stress in Information System Professionals? Communications of the ACM, 47, 99-102.http://dx.doi.org/10.1145/971617.971623

25. Shapley, L. (2013). Most Stressful Jobs: Journalism Careers Make 2013 List. The Denver Post, 16 April 2013. http://blogs.denverpost. com/editors/2013/04/16/media-jobs-make-list-of-top-stressful-careers-of-2013/857/

26. Sidle, S. (2008). Workplace Stress Management Interventions: What Works Best? Academy of Management Perspectives, 22, 111-112.http://dx.doi.org/10.5465/AMP.2008.34587999

27. Tarkovsky, S. (2007). Professional Stress—All You Need to Know to Beat It.http://EzineArticles.com/434206

28. Teasdale, E. L. (2006). Workplace Stress. Psychiatry, 5, 251-254.http://dx.doi.org/10.1053/j.mppsy.2006.04.006

29. Turnage, J. J., & Spielberger, C. D. (1991). Job Stress in Managers, Professionals, and Clerical Workers. Work & Stress, 5, 165-176.http://dx.doi.org/10.1080/02678379108257015

30. Tyler, K. (2006). Restructuring Policies and Workloads, Along with Providing Training and Support Services, Can Help Reduce Employee Stress. HR Magazines, 51.

31. Uehata, T. (1991). Long Working Hours and Occupational Stress-Related Cardiovascular Attacks among Middle-Aged Workers in Japan. Journal of Human Ergology, 20, 147-153.

32. Ulrich, C., O'Donnell, P., Taylor, C., Farrar, A., Danis, M., & Grady, C. (2007). Ethical Climate, Ethics Stress, and the Job Satisfaction of Nurses and Social Workers in the United States. Social Science & Medicine, 65, 1708-1719.http://dx.doi.org/10.1016/j.socscimed.2007.05.050

33. University of Cambridge (2014). Causes of Work-Related Stress.http://www.admin.cam.ac.uk/offices/hr/policy/stress/causes.html

34. World Health Organization (1948). Preamble to the Constitution of the World Health Organization as Adopted by the International Health Conference, New York, 19-22 June 1946; Signed on 22 July 1946 by the Representatives of 61 States (Official Records of the World Health Organization No. 2, pp. 100) and Entered into Force on 7 April 1948. The Definition Has Not Been Amended since 1948.

35. Zakaria, M., & Azad, A. K. (2009). Journalism as a Profession in Bangladesh: An Overview. The Chittagong University Journal of Arts and Humanities, XXII, 236-248.

Chapter 10

EXTRACTING ASSOCIATION PATTERNS IN NETWORK COMMUNICATIONS

Javier Portela [1], Luis Javier García Villalba [1], Alejandra Guadalupe Silva Trujillo [1,2], Ana Lucila Sandoval Orozco [1] and Tai-hoon Kim [3]

[1]Group of Analysis, Security and Systems (GASS), Department of Software Engineering and Artificial Intelligence (DISIA), Faculty of Information Technology and Computer Science, Office 431, Universidad Complutense de Madrid (UCM), Calle Profesor José García Santesmases, 9, Ciudad Universitaria, Madrid 28040, Spain

[2]Facultad de Ingeniería, Universidad Autónoma de San Luis Potosí (UASLP), Zona Universitaria Poniente, San Luis Potosí 78290, Mexico

[3]Department of Convergence Security, Sungshin Women's University, 249-1 Dongseon-dong 3-ga, Seoul 136-742, Korea

ABSTRACT

In network communications, mixes provide protection against observers hiding the appearance of messages, patterns, length and links between senders and receivers. Statistical disclosure attacks aim to reveal the identity of senders and receivers in a communication network setting when it is protected by standard techniques based on mixes. This work aims to develop a global statistical disclosure attack to detect relationships between users. The only information used by the attacker is the number of messages sent and received by each user for each round, the batch of messages grouped by the anonymity system. A new modeling framework based on contingency tables is used. The assumptions are more flexible than those used in the literature, allowing to apply the method to multiple situations automatically, such as email data or social networks data. A classification scheme based on combinatoric solutions of the space of rounds retrieved is developed. Solutions about relationships between users are provided for all pairs of users simultaneously, since the dependence of the data retrieved needs to be addressed in a global sense.

INTRODUCTION

When information is transmitted through the Internet, it is typically encrypted in order to prevent others from being able to view it. The encryption can be successful, meaning that the keys cannot be easily guessed within a very long period of time. Even if the data themselves are hidden, other types of information may be vulnerable. In the e-mail framework, anonymity concerns the senders "identity, receivers" identity, the links between senders and receivers, the protocols used, the size of data sent, timings, etc. Since [1] presented the basic ideas of the anonymous communications systems, researchers have developed many mix-based and other anonymity systems for different applications, and attacks on these systems have also been developed. Our work aims to develop a global statistical attack to disclose relationships between users in a network based on a single mix anonymity system.

INTRODUCING ANONYMOUS COMMUNICATIONS

The infrastructure of the Internet was initially planned and developed to be an anonymous channel, but nowadays, it is well known that anybody can spy on it with different non-robust tools, like, for example, using sniffers and spoofing techniques. Since the Internet's proliferation and the use of some services associated with it, such as web searchers, social networks, webmail and others, privacy has become a very important research area, not just for security IT experts or enterprises. Connectivity and the enormous flow of information available on the Internet are a very powerful tool to provide knowledge and to implement security measures to protect systems.

Anonymity is a legitimate means in many applications, such as web browsing, e-vote, e-bank, e-commerce and others. Popular anonymity systems are used by hundreds of thousands people, such as journalists, whistle blowers, dissidents and others. It is well known that encryption does not guarantee the anonymity required for all participants. Attackers can identify traffic patterns to deduce who, when and how often users are in communication. The communication layer is exposed to traffic analysis, so it is necessary to anonymize it, as well as the application layer that supports anonymous cash, anonymous credentials and elections.

Anonymity systems provide mechanisms to enhance user privacy and to protect computer systems. Research in this area focuses on developing, analyzing and executing anonymous communication networks attacks.

Two categories for anonymous communication systems are commented on below: high latency systems and low latency systems. Both systems are based on Chaum's proposal [1] that introduced the concept of mixing.

High latency anonymity systems aim to provide a strong level of anonymity and are oriented to limited activity systems that do not demand quick responses, such as email systems. These systems are message-oriented systems.

Low latency anonymity systems can be used for interactive traffic, for example web applications, instant messaging and others. These systems are connection-based systems and are used to defend from a partial attacker who can compromise or observe just a part of the system. According to its nature, these systems are more susceptible to timing attacks and traffic analysis attacks. The majority of these systems depend on onion routing [2] for anonymous communication.

In low latency communication systems, an attacker only needs to observe the flow of the data stream to link sender and receptor users. Traditionally, in order to prevent this attack, dummy packets are added and delays incorporated into stream data to make the traffic between users uniform. The previously mentioned scenario can be useful for passive attackers that do not insert timing partners into the traffic to compromise anonymity. An active attacker can control routers of the network. Timing attacks are one of the main challenges in low latency anonymous communication systems. These attacks are closely related to traffic analysis in mix networks.

Traffic analysis techniques belong to the family of methods to infer information from the patterns in a communication system. Even when communication content has been ciphered, information routing needs to be sent clearly for routers to know the next package's destination in the network. Every data packet traveling on the Internet contains the node addresses of sending and recipient nodes. Therefore, it is well understood that, actually, no packet can be anonymous at this level.

Mixes and the Mix Network Model

Mixes are considered the base for building high latency anonymous communication systems. In network communications, mixes provide protection against observers hiding the appearance of messages, patterns, length and links between senders and receivers. Chaum [1] introduced also the concept of anonymous email. Their model suggested hiding the correspondence between senders and receivers encrypting messages and reordering them through a path of mixes before relaying them to their destinations. The set of the most likely receivers is calculated for each message in the sequence, and intersection of sets will make it possible to know who the receiver of the stream is.

A mix networks aims to hide the correspondences between the items in its input and those in its output, changing the incoming packets appearance through cryptographic operations (see Figure 1). The anonymity set is the set of all possible entities who might execute an action. The initial process in order for Alice send a message to Bob using a mix system is to prepare the message. The first phase is to choose the message transmission path; it has a specific order for iteratively sending messages before arriving at its final destination. It is recommended to use more than one mix in every path for improving system security. The next phase is to utilize the public keys of the chosen mixes for encrypting the message in the inverse order that they were chosen. Therefore, the public key of the last mix initially encrypts the message, then the next one before the last one, and finally, the public key of the first mix will be used. Every time a message is encrypted, a layer is built, and the next node address is included. This way, when the first mix gets a message prepared, this will be decrypted with its correspondent private key and will get the next node address.

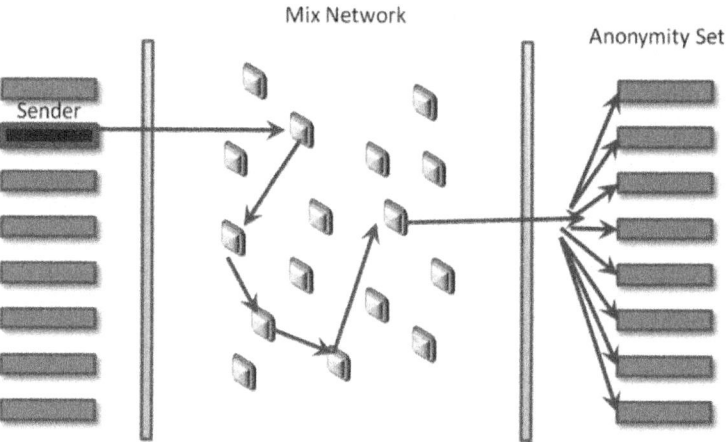

Figure 1: Mix network model.

An observer or an active attacker should not be able to find the link between the bit pattern of encoded messages arriving at the mix and decoded messages departing from it. Appending a block of random bits at the end of the message has the purpose of making messages uniform in size.

ISDN-Mixes

The first proposal for the practical application of mixes [3] showed the way a mix-net could be used with ISDN lines to anonymize a telephone user's real location. The origin of this method took into account the fact that mixes in their original form imply a significant data expansion and significant delays, and therefore, it was often considered infeasible to apply them to services with higher bandwidth and real-time requirements. The protocol tries to defeat these problems.

Remailers

The first Internet anonymous remailer was developed in Finland and was very simple to use. A user added an extra header to the e-mail pointing out its final destination: an email address or a Usenet newsgroup. A server receives messages with embedded instructions about where to send them next without revealing their origin. All standard-based email messages include the source and transmitting entities at the headers. The full headers are usually eliminated. The application replaces the original email's source address with the remailer's address.

Babel [4], Mixmaster [5] and Mixminion [6] are some others anonymous communication designs. The differences between systems will not be addressed in our work. We centered only on senders and receivers active in a period of time, and we do not take into account message reordering, because this does not affect our attack. Onion routing [2] is another design used to provide low latency connection for web browsing and other interactive services. It is important to specify that our method does not address this kind of design; they can be treated by short-term timing or packet counting attacks [7].

THE FAMILY OF MIX SYSTEMS ATTACKS

The attacks against mix systems are intersection attacks and aim to reduce the anonymity by linking senders with the messages that they send, receivers with the messages that they receive or linking senders with receivers. Attackers can derive relations of frequency through observation of the network, compromising mixes or keys, delaying or altering messages. They can deduce the messages' most probable destinations through the use of false messages sent to the network and using this technique to isolate target messages and to derive their properties. Traffic analysis belongs to a family of techniques used to deduce pattern information in a communication system. It has been proven that cipher by itself does not guarantee anonymity. See [8] for a review of traffic analysis attacks.

The Disclosure Attack

In [9], Agrawal and Kesdogan presented the disclosure attack, an attack centered on a single batch mix, aiming to retrieve information from a particular sender, called Alice. The attack is global, in the sense that it retrieves information about the number of messages sent by Alice and received by other users, and passive, in the sense that attackers cannot alter the network, for example, by sending false messages or delaying existent messages.

It is assumed that Alice has exactly m recipients and that Alice sends messages with some probability distribution to each of her recipients; also that she sends exactly one message in each batch of b messages. The attack is modeled considering a bipartite graph G. Through numerical algorithms, disjoint sets of recipients would be identified to reach, through intersection, the identification of Alice recipients. The authors use several strategies in order to estimate the average number of observations for achieving the disclosure attack. The assumptions are: (i) Alice participates in all batches; and (ii) only one of Alice's peer partners is in the recipient set of all batches. This attack is computationally expensive, because it takes an exponential time analyzing the number of messages to identify a mutually disjoint set of recipients. The main bottleneck for the attacker derives from an NP-complete problem when it is applied to big networks. The authors claim the method performs well on very small networks.

Statistical Disclosure Attacks

In [10], Danezis presents the statistical disclosure attack, maintaining some of the assumptions made in [9]. In the statistical disclosure attack, recipients are ordered in terms of probability. Alice must demonstrate consistent behavior patterns in the long term to obtain good results. The Statistical Disclosure Attack (SDA) requires less computational effort by the attacker and gets the same results. The method tries to reveal the most likely set of Alice's friends using statistical operations and approximations.

Statistical disclosure attacks when threshold mixing or pool mixing are used are treated also in [11], maintaining the assumptions of precedent articles, that is, focusing on one user, Alice, and supposing that the number of recipients of Alice is known. Besides, the threshold parameter B is also supposed to be known. One of the main characteristics of intersection attacks counts on a fairly consistent sending pattern or a specific behavior of anonymous network users.

Mathewson and Dingledine in [12] make an extension of the original SDA. One of the more significant differences is that they regard real social networks to have scale-free network behavior and also consider that such behavior changes slowly over time. The results show that increasing message variability makes the attack slow by increasing the number of output messages; assuming all senders choose with the same probability all mixes as entry and exit points and the attacker is a partial observer of the mixes. Two-sided statistical disclosure attacks [13] use the possibilities of replies between users to make the attack stronger. This attack assumes a more realistic scenario, taking into account the user behavior on an email system. Its aim is to estimate the distribution of contacts of Alice and to deduce the receivers of all of the messages sent by her. The model considers N as the number of users in the system that send and receive messages. Each user n has a probability distribution Dn of sending a message to other users. At first, the target, Alice, is the only user that will be modeled as replying to messages with a probability r. An inconvenient detail for applications on real data is the assumption that all users have the same number of friends and send messages with uniform probability.

Perfect matching disclosure attacks [14] try to use simultaneous information about all users to obtain better results related to the disclosing of the Alice set of recipients. This attack is based on graph theory, and it does not consider the possibility that users send messages with different frequencies. An extension proposal considers a normalized SDA.

Danezis and Troncoso [15] present a new modeling approach, called Vida, for anonymous communication systems. These are modeled probabilistically, and Bayesian inference is applied to extract patterns of communications and user profiles. The authors developed a model to represent long-term attacks against anonymity systems. Assume each user has a sending profile, sampled when a message is to be sent to determine the most likely receiver. Their proposal includes: (1) the Vida black-box model representing long-term attacks against any anonymity systems. Bayesian techniques are used to select the candidate sender of each message: the sender with the highest a posterioriprobability is chosen as the best candidate. The evaluation includes a very specific scenario considering the same number of senders and receivers. Each sender is assigned to five contacts randomly, and everyone sends messages with the same probability.

In [16], a new method to improve the statistical disclosure attack, called the hitting set attack, is introduced. Frequency analysis is used to enhance the applicability of the attack, and duality checking algorithms are also used to resolve the problem of improving the space of solutions. Mallesh and Wright [17] introduces the reverse statistical disclosure attack. This attack uses observations of all users sending patterns to estimate both the targeted user's sending pattern and her receiving pattern. The estimated patterns are combined to find a set of the targeted user's most likely contacts.

In [18], an extension to the statistical disclosure attack, called SDA-2H, is presented, considering the situation where cover traffic, in the form of fake or dummy messages, is employed as a defense.

Perez-Gonzalez et al. [19] presents a least squares approximation to the SDA, to recover users' profiles in the context of pool mixes. The attack estimates the communication user partners in a mix network. The aim is to estimate the probability of Alice sending a message to Bob; this will derive sender and receiver profiles applicable for all users. The assumptions are: the probability of sending a message from a user to a specific receiver is independent of previous messages; the behavior of all users are independent from one other; any incoming message in the mix is considered a priori sent by any user with a uniform probability; and the parameters used to model the statistical behavior do not change over time.

In [20], a timed binomial pool mix is used, and two privacy criteria to develop dummy traffic strategies are taken into account: (i) increasing the estimation error for all relationships by a constant factor; and (ii) guaranteeing a minimum estimation error for any relationship. The model consists of a set of N senders exchanging messages with a set of M receivers. To simulate the system, consider the same number of senders and receivers and assume users send messages with the same probability. Other work also based on dummy or cover traffic is presented in [21]. This assumes users are not permanently online so, so they cannot send cover traffic uniformly. They introduce a method to reveal Alice›s contacts with high probability, addressing two techniques: sending dummy traffic and increasing random delays for messages in the system.

Each one of the previous works has assumed very specific scenarios, but none of them solves the problems that are presented by real-world data. In order to develop an effective attack, the special properties of network human communications must be taken into account. Researchers have hypothesized that some of these attacks can be extremely effective in many real-world contexts. Nevertheless, it is still an open problem under which circumstances and for how long of an observation these attacks would be successful.

FRAMEWORK AND ASSUMPTIONS

This work addresses the problem of retrieving information about relationships or communications between users in a network system, where partial information is obtained. The information used is the number of messages sent and received by each user. This information is obtained in rounds that can be determined by equally-sized batches of messages, in the context of a threshold mix, or alternatively by equal length intervals of time, in the case that the mix method consists of keeping all of the messages retrieved at each time interval and then relaying them to their receivers, randomly reordered.

- The basic framework and assumptions needed to develop our method are the following:

- The attacker knows the number of messages sent and received by each user in each round.

- The round can be determined by the system (batches) in a threshold mix context or can be based on regular intervals of time, where the attacker gets the aggregated information about messages sent and received, in the case of a timed mix, where all messages are reordered and sent each period of time.

- The method is restricted, at this moment, to threshold mixing with a fixed batch size or, alternatively, to a timed mix, where all messages received in a fixed time period are relayed randomly, reordered with respect to their receivers.

- No restriction is made from before about the number of friends any user has nor about the distribution of messages sent. Both are considered unknown.

- The attacker controls all users in the system. In our real data application, we aim at all email users of a domain sent and received within this domain.

The method introduced in this work allows one to address these general settings in order to derive conclusions about the relationships between users. Contrary to other methods in the literature, there are no restrictions about user relationships (number of friends, distribution of messages), and therefore, it can be used in a wider context. Furthermore, our proposition is new in the methodological sense: this is a novel approach to the problem, by means of contingency table setting and the extraction of solutions by sampling.

In an email context, this attack can be used if the attacker has access, at regular time intervals, to the information represented by the number of messages received and the number of messages sent for each user, in a closed domain or intranet, where all users are controlled. This situation can also be extended to mobile communications or social networks and could be used, for example, in the framework of police communication investigations.

MARGINAL INFORMATION AND FEASIBLE TABLES

The attacker obtains, in each round, information about how many messages each user sends and receives. Usually, the sender and receiver set is not the same, even if some users are senders and also receivers in some rounds. Furthermore, the total number of users of the system N is not present in each round, since only a fraction of them are sending or receiving messages. Figure 2 represents a round with only six users.

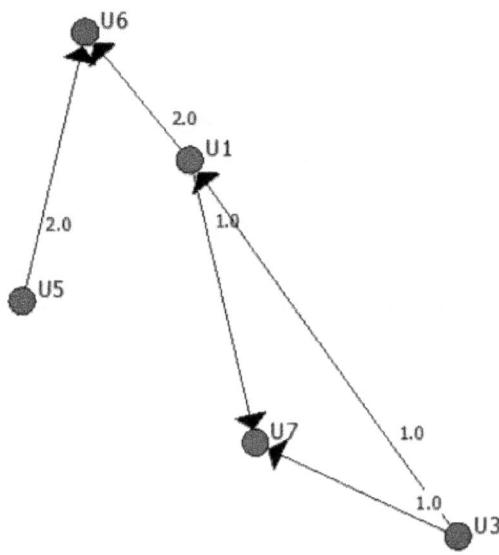

Figure 2: Graphical representation of one round.

The information of this round can be represented in a contingency table (see Table 1), where the element (i, j) represents the number of messages sent from user i to user j:

Table 1: Example of a contingency table

Senders\Receivers	U1	U6	U7	Total Sent
U1	0	2	1	3
U3	1	0	1	2
U5	0	2	0	2
Total received	1	4	2	7

The attacker only sees the information present in the aggregated marginals, which means, in rows, the number of messages sent by each user, and in columns, the number of messages received by each user. In our example, only the sending pairs of vectors (U1 U3 U5) (3 2 2) and receiver pairs of vectors (U1 U6 U7) (1 4 2) are known.

There are many possible tables that can lead to the table with the given marginals that the attacker is seeing, making it impossible, in most cases, to derive direct conclusions about relationships. The feasible space of the tables' solution of the integer programming problem can be very large. In the example, there are only 16 possible different solutions and only one true solution.

Solutions (feasible tables) can be obtained via algorithms, such as the branch and bound algorithm or other integer programming algorithms. In general they do not guarantee covering evenly all possible tables/solutions, since they are primarily designed to converge to one solution. The simulation framework presented in this article allows us to obtain a large quantity of feasible tables (in the most problematic rounds, it takes approximately three minutes to obtain one million feasible tables). In many of the rounds with a moderate batch size, all feasible tables are obtained.

An algorithm that takes into account the information contained over all of the rounds retrieved is developed in the next section.

STATISTICAL DISCLOSURE ATTACK BASED ON PARTIAL INFORMATION

The main objective of the algorithm we propose is to derive relevant information about the relationship (or not) between each pair of users. The information obtained by the attacker is the marginal sums, by rows and columns, of each of the rounds 1, ..., T, where T is the total number of rounds. Note that in each round, the dimension of the table is different, since we do not take into account users that are not senders (row marginal = 0), nor users that are not receivers (column marginal = 0). We say element (i, j) is "present" at one round if the i and j corresponding marginals are not zero. That means that user i is present in this round as the sender and user j is present as the receiver.

A final aggregated matrix A can be built, summing up all of the rounds and obtaining a table with all messages sent and received from each user for the whole time interval considered for the attack. Each element (i, j) of this final table would represent the number of messages sent by i to j in total. Although the information obtained in each round is more precise and relevant (because of the lower dimension and combinatoric possibilities), an accurate estimate of the final table is the principal objective, because a zero in elements (i, j) and (j, i) would mean no relationship between these users (no messages sent from i to j nor from j to i). A positive number in an element of the estimated final table would mean that some message is sent in some round, while a zero would mean no messages are sent in any round, that is, no relationship.

We consider all rounds as independent events. The first step is to obtain the higher number of feasible tables that is possible for each round, taking into account time restrictions. This will be the basis of our attack. In order to obtain feasible tables we use Algorithm 1, based on [22]. It consists of filling the table column by column and computing the new bounds for each element before it is generated.

Algorithm 1

① Begin with column one, row one:

Generate n_{11} from an integer uniform distribution in the bounds according to Equation (1), where $i = 1, j = 1$.

Let r be the number of rows.

② For each row element n_{k1} in this column, if row elements until $k-1$ have been obtained, new bounds for n_{k1} are according to Equation (1):

$$\max\left(0, (n_{+1} - \sum_{i=1}^{k-1} n_{i1}) - \sum_{i=k+1}^{r} n_{i+}\right) \leq n_{k1} \leq \min\left(n_{k+}, n_{+1} - \sum_{i=1}^{k-1} n_{i1}\right) \tag{1}$$

The element n_{k1} is then generated by an integer uniform in the fixed bounds.

③ The last row element is automatically filled, since the lower and upper bounds coincide, letting $n_{(k+1)+} = 0$ by convenience.

④ Once this first column is filled, the row margins n_{i+} and total count n are actualized by subtraction of the already fixed elements, and the rest of the table is treated as a new table with one less column.

The algorithm fixes column by column until the whole table is filled.

The time employed depends on the complexity of the problem (number of elements, mean number of messages). In our email data, even for a large number of elements, this has not been a problem. For large table sizes in our applications, it takes approximately 3 min to obtain one million feasible tables in rounds with 100 cells and 10 on a PC with Intel processor 2.3 GHz and 2 GB RAM.

Repeating the algorithm as it is written for each generated table does not lead to uniform solutions, that is some tables are more probable than others due to the order used when filling columns and rows. Since we must consider a priori all solutions for a determined round equally possible, two further modifications are made: (i) random reordering of rows and columns before a table is generated; and (ii) once all tables are generated, only distinct tables are kept to make inferences. These two modifications have resulted in an important improvement of the performance of our attack, lowering the mean misclassification rate to about a 20% in our simulation framework.

Deciding the number of tables to be generated poses an interesting problem. Computing the number of distinct feasible tables for a contingency table with fixed marginals is still an open problem that has been addressed via algebraic methods [23] and by asymptotic approximations [24], but in our our case, the margin totals are small and depend on the batch size; therefore, it is not guaranteed that asymptotic approximations hold. The best approximation so far to count the feasible tables is to use the generated tables.

Chen et al. [22] show that an estimate of the number of tables can be obtained by averaging over all of the generated tables the value $\frac{1}{q(T)}$ according to the Algorithm 2.

Algorithm 2

---

① $q(T)$ is the probability of obtaining the table T and is computed iteratively, imitating the simulation process according to Equation (2).

② $q(t_1)$ is the probability of the actual values obtained for Column 1, obtained by multiplying the uniform probability for each row element in its bounds. $q(t_2 \mid t_1)$ and subsequent terms are obtained in the same way, within the new bounds restricted to the precedent columns fixed values:

$$q(T) = q(t_1)q(t_2 \mid t_1)q(t_3 \mid t_1, t_2)...q(t_c \mid t_1, t_2, ..., t_{c-1}) \qquad (2)$$

The number of feasible tables goes from moderate values, such as 100,000, that can be easily addressed, getting all possible tables via simulation, to very high numbers, such as 10^{13}. Generating all possible tables for this last example would take, with the computer we are using, a Windows 7 PC with 2.3 GHz and 4 GB RAM, at least 51 days. The quantity of feasible tables is the main reason why it is difficult for any deterministic intersection-type attack to work, even with low or moderate user dimensions. Statistical attacks need to consider the relationships between all users to be efficient, because the space of solutions for any individual user is dependent on all other users' marginals. Exact trivial solutions can be, however, found at some time in the long run, if a large number of rounds are obtained.

In our setting, we try to obtain the largest number of tables that we can, given our time restrictions, obtaining a previous estimate of the number of feasible tables and fixing the highest number of tables that can be obtained for the most problematic rounds. However, an important issue is that once a somewhat large number of tables is obtained, good solutions depend more on the number of rounds treated (time horizon or total number of batches considered) than on generating more tables. In our simulations, there is generally a performance plateau in the curve that represents the misclassification rate versus the number of tables generated, since a sufficiently high number of tables is reached. This minimum number of tables to be generated depends on the complexity of the application framework.

The final information obtained consists of a fixed number of generated feasible tables for each round. In order to obtain relevant information about relationships, there is a need to fix the most probable zero elements. For each element, the sample likelihood function at zero $f\square$ $(X \mid p_{ij} = 0)$ is estimated. This is done by computing the percent of tables with that element being zero in each round that the element is present and multiplying the estimated likelihood obtained in all of these rounds (the element will be zero for the final table if it is zero for all rounds). If we are estimating the likelihood for the element (i, j) and are generating M tables per round, we use the following expressions:

$n_t^{(i,j)}$ the number of tables with element $(i, j) = 0$ in round t.

$N_{present}$ = the number of rounds with element (i, j) present.

X = the sample data, given by marginal counts for each round.

$$log(\hat{f}(X \mid p_{ij} = 0)) = -N_{present} \log(M) + \sum_{t=1,(i,j)present}^{T} \log(n_t^{(i,j)})$$

(3)

Final table elements are then ordered by the estimated likelihood at zero, with the exception of elements that were already trivial zeros (elements that represent pair of users that have never been present at any round).

Elements with the lowest likelihood are then considered candidates to insert as a "relationship". The main objective of the method is to detect accurately:

1. cells that are zero with a high likelihood (no relationship i → j);

2. cells that are positive with high likelihood (relationship i → j).

In our settings the likelihood values at $p_{ij} = 0$ are bounded in the interval [0, 1]. Once these elements are ordered by most likely to be zero to less, a classification method can be derived based on this measure. A theoretical justification of the consistency of the ordering method is given below.

Proposition 1

Let us consider, a priori, that for any given round k, all feasible tables, given the marginals, are equiprobable.

Let p_{ij} be the probability of element (i, j) being zero at the final matrix A, which is the aggregated matrix of sent and received messages over all rounds. Then, the product of the proportion of feasible tables with $x_{ij} = 0$ at each round, Q^{ij} leads to an ordering between elements, such that if $Q^{ij} > Q^{i'j'}$; then, the likelihood of data for $p_{ij} = 0$ is bigger than the likelihood of data for $p_{i'j'} = 0$.

Proof

If all feasible tables for round k are equiprobable, the probability of any feasible table is $p_k = \frac{1}{\#[X]_k}$, where $\#[X]_k$ is the total number of feasible tables in round k.

For elements with $p_{ij} = 0$, it is necessary that $x_{ij} = 0$ for any feasible table. The likelihood for $p_{ij} = 0$ is then:

$$P([X]_k \mid p_{ij} = 0) = \frac{\#[X \mid x_{ij} = 0]_k}{\#[X]_k}$$

where $\#[X \mid x_{ij} = 0]_k$ denotes the number of feasible tables with the element $x_{ij} = 0$.

Let k = 1, …, t independent rounds. The likelihood at $p_{ij} = 0$, considering all rounds, is:

$$Q^{ij} = \prod_{k=1}^{t} P([X]_k \mid p_{ij} = 0) = \prod_{k=1}^{t} \frac{\#[X \mid x_{ij} = 0]_k}{\#[X]_k}$$

and the log likelihood:

$$\log(Q^{ij}) = \sum_{k=1}^{t} \log(\#[X \mid x_{ij} = 0]_k) - \sum_{k=1}^{t} \log(\#[X]_k)$$

Then, the proportion of elements with $x_{ij} = 0$ at each round leads to an ordering between elements, such that if $Q^{ij} > Q^{i'j'}$, then the likelihood of data for $p_{ij} = 0$ is bigger than the likelihood of data for $p_{i'j'} = 0$.

Our method is not based on all of the table solutions, but on a consistent estimator of Q^{ij}. For simplicity, let us consider a fixed number of M sampled tables at every round.

Proposition 2

Let $[X]_k^1, …, [X]_k^M$ be a random sample of size M of the total $\#[X]_k$ of feasible tables for round k. Let $w_k^{(i,j)} = \frac{\#[X|x_{ij}=0]_k^M}{M}$ be the sample proportion of feasible tables with $x_{ij} = 0$ at round k. Then, the statistic $q^{ij} = \prod_{k=1}^{t} \frac{\#[X|x_{ij}=0]_k^M}{M}$ is such that, for any pair of elements (i, j) and (i', j'), $q^{ij} > q^{i'j'}$ implies, in convergence, a higher likelihood for $p_{ij} = 0$ than for $p_{i'j'} = 0$.

Proof. (1) Let $\#[X]_k$ be the number of feasible tables at round k. Let $[X]_k^1, …, [X]_k^M$ be a random sample of size M of the total $\#[X]_k$. Random reordering of columns and rows in Algorithm 1, together with the elimination of equal tables, assures that it is a random sample. Let $\#[X \mid x_{ij} = 0]_k^M$ be the number of sample tables with element $x_{ij} = 0$. Then, the proportion $w_k^{(i,j)} = \frac{\#[X|x_{ij}=0]_k^M}{M}$ is a consistent and unbiased estimator of the true proportion $W_k^{(i,j)} = \frac{\#[X|x_{ij}=0]_k}{\#[X]_k}$. This is a known result from finite population sampling. As $M \to \#[X]_k$, $w_k^{(i,j)} \to W_k^{(i,j)}$.

(2) Let $k = 1, …, t$ independent rounds. Then, given a sample of proportion estimators $w_1^{(i,j)}, …, w_t^{(i,j)}$ of $W_1^{(i,j)}, …, W_t^{(i,j)}$, consider the function

$$f(w_1^{(i,j)}, …, w_t^{(i,j)}) = \sum_{k=1}^{t} \log(w_k^{(i,j)}) \text{ and } f(W_1^{(i,j)}, …, W_t^{(i,j)}) = \sum_{k=1}^{t} \log(W_k^{(i,j)}).$$

Given the almost sure convergence of each $w_k^{(i,j)}$ to each $W_k^{(i,j)}$ and the continuity of the logarithm and sum functions, the continuous mapping theorem assures convergence in probability, $f(w_1^{(i,j)}, …, w_t^{(i,j)}) \xrightarrow{P} f(W_1^{(i,j)}, …, W_t^{(i,j)})$. Then, $\log(q^{ij}) = f(w_1^{(i,j)}, …, w_t^{(i,j)})$ converges to $log(Q^{ij}) = f(W_1^{(i,j)}, ..., W_t^{(i,j)})$. Since the exponential function is continuous and monotonically increasing, applying the exponential function to both sides leads to the convergence of q^{ij} to Q^{ij}, so that $q^{ij} > q^{i'j'}$ implies, in convergence, $Q^{ij} > Q^{i'j'}$ and, then, higher likelihood for $p_{ij} = 0$ than for $p_{i'j'} = 0$.

Given all pairs of senders and receivers (i, j) ordered by the statistic q^{ij}, it is necessary to select a cut point in order to complete the classification scheme and to decide whether a pair communicates ($p_{ij} > 0$) or not ($p_{ij} = 0$). That is, it is needed to establish a value c, such that $q^{ij} > c$ implies $p_{ij} = 0$ and $q^{ij} \le c$ implies $p_{ij} > 0$. The defined statistic q^{ij} is bounded in [0, 1], but this is not strictly a probability, so fixing a priori a cut-point, such as 0.5, is not an issue. Instead, there are some approaches that can be used:

In some contexts (email, social networks), the proportion of pairs of users that communicate is approximately known . This information can be used to select the cut point from the ordering. That is, if about 20% of pairs of users are known to communicate, the classifier would give a value "0" (no communication) to the upper 80% elements (i, j), ordered by the statistic q^{ij}, and a value "1" (communication) to the lower 20% of elements.

If the proportion of zeros is unknown, it can be estimated, using the algorithm for obtaining feasible tables over the known marginals of the matrix A and estimating the proportion of zeros by the mean proportion of zeros over all of the simulated feasible tables.

PERFORMANCE OF THE ATTACK

In this section, simulations are used to study the performance of the attack.

Each element (i, j) of the matrix A can be zero (no communication) or strictly positive. The percentage of zeroes in this matrix is a parameter, set a priori to observe its influence. In a closed-center email communications, this number can be between 70% and 99% . However, intervals from 0.1 (high communication density) to 0.9 (low communication density) are used here for different practical situations. Once this percentage is set, a randomly chosen percent of elements are set to zero and then are zero for all of the rounds.

The mean number of messages per round for each positive element (i, j) is also set a priori. This number is related, in practice, to the batch size that the attacker can obtain. As the batch size (or time length interval of the attack) decreases, the mean number of messages per round decreases, making the attack more efficient.

Once the mean number of messages per round is determined for each positive element (λ_{ij}), a Poisson distribution with mean λ_{ij}, $P(\lambda_{ij})$, is used to generate the number of messages for each element, for each of the rounds. External factors, given by the context (email, social networks, etc.) that have an effect on the performance of the method are monitored to observe their influence:

The number of users: In a network communication context with N users, there exist N potential senders and N receivers in total, so that the maximum dimension of the aggregated matrix A is N^2. As the number of users increases, the complexity of round tables and the number of feasible tables increases, so that it could negatively affect the performance of the attack.

The percent of zero elements in the matrix A: These zero elements represent no communication between users. As will be seen, this influences the performance of the method.

The mean frequency of messages per round for positive elements: This is directly related to the batch size, and when it increases, the performance is supposed to be affected negatively.

The number of rounds: As the number of rounds increases, this is supposed to improve the performance of the attack, since more information is available. One factor related to the settings of the attack method is also studied.

The number of feasible tables generated by round: This affects computing time, and it is necessary to study to what extent it is useful to obtain too many tables. This number can be variable, depending on the estimated number of feasible tables for each round .

The algorithm results in a binary classification, where zero in an element (i, j) means no relationship of sender-receiver from i to j and one means a positive relationship of sender-receiver.

Characteristic measures for binary classification tests include the sensitivity, specificity, positive predictive value and negative predictive value. Letting TP be true positives, FP false positives, TN true negatives and FN false negatives:

Sensitivity= $\frac{TN}{TN+FP}$ measures the capacity of the test to recognize true negatives.

Specificity= $\frac{TP}{TP+FN}$ measures the capacity of the test to recognize true positives.

Positive predictive value = $\frac{TP}{TP+FP}$ measures the precision of the test to predict positive values.

Negative predictive value = $\frac{TN}{TN+FN}$ measures the precision of the test to predict positive values.

Classification rate $= \frac{TN+TP}{TN+TP+FN+FP}$ measures the percent of elements well classified.

Figures 3 and 4 show the simulation results. When it is not declared, values of $p_0 = 0.7$, $\lambda = 2$, $N = 50$ users and the number of rounds $= 100$ are used as base values.

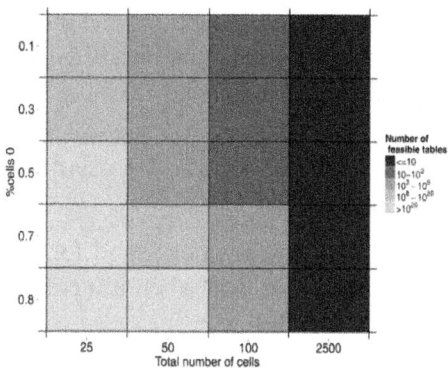

Figure 3: Number of feasible tables per round, depending on % of cells of zero and the total number of cells.

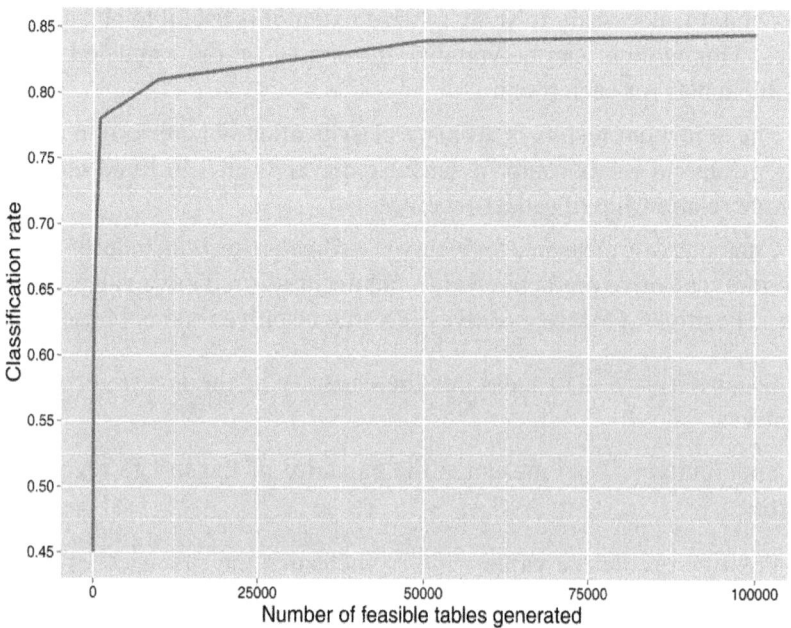

Figure 4: Classification rate as function of the number of feasible tables generated per round.

Figure 3 shows that as the number of cells (N^2, where N is the number of users) increases and the percent of cells that are zero decreases, the number of feasible tables per round increases. For a moderate number of users, such as 50, the number of feasible tables is already very high, greater than 10^{20}. This does not have a strong effect on the main results, except for lower values. As can be seen in Figure 4, once a sufficiently high number of tables per round is generated, increasing this number does not lead to significant improvement of the correct classification rate.

Figure 5 shows that the minimum classification rate is attained at a percent of cells of zero (users that do not communicate) near 0.5. As this percent increases, the true positive rate decreases, and the true negative rate increases.

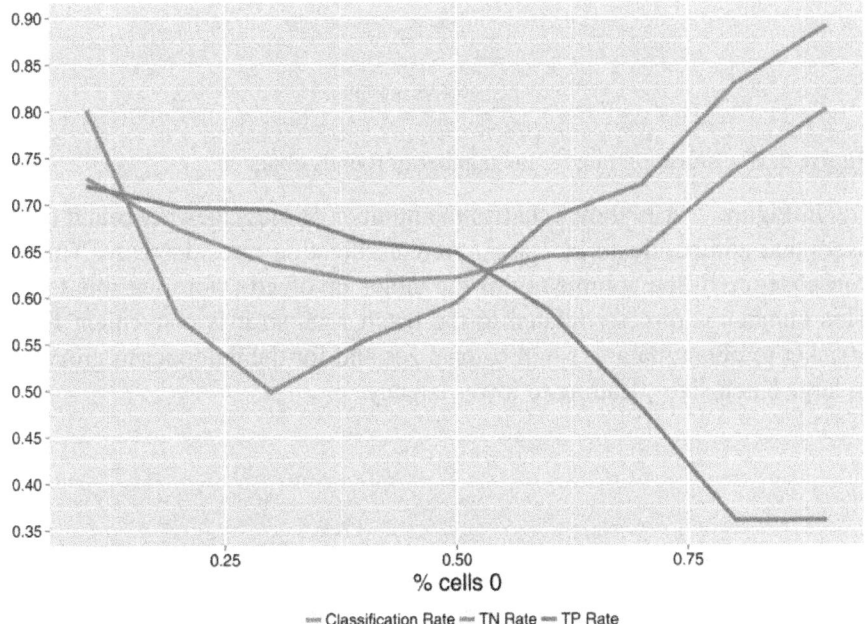

Figure 5: Classification rate, true positive rate and true negative ratevs. the percent of cells of zero.

As the attacker gets more information, that is more rounds are retrieved, the classification rate gets better. Once a high number of rounds is obtained, there is no further significant improvement, as is shown in Figure 6.

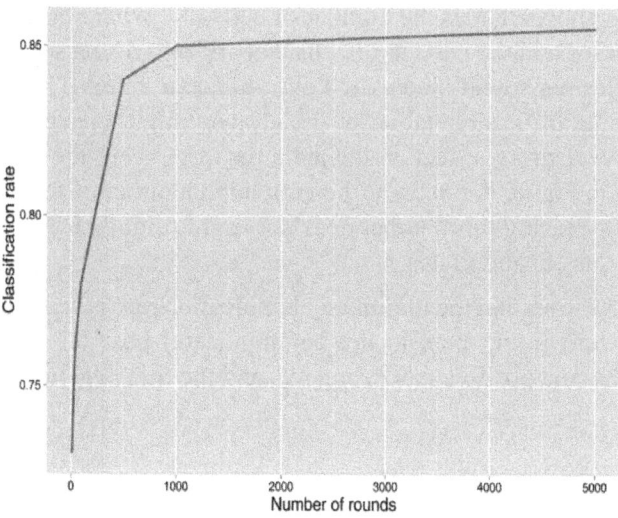

Figure 6: Classification rate vs. the number of rounds obtained.

In Figure 7, it is shown that as the number of messages per round (λ) for users that communicate increases, the classification rates decrease. This is a consequence of the complexity of the tables involved (more feasible tables). This number is directly related to the batch size, so it is convenient for the attacker to obtain data in small batch sizes and for the defender to group data in large batch sizes, leading to lower latency.

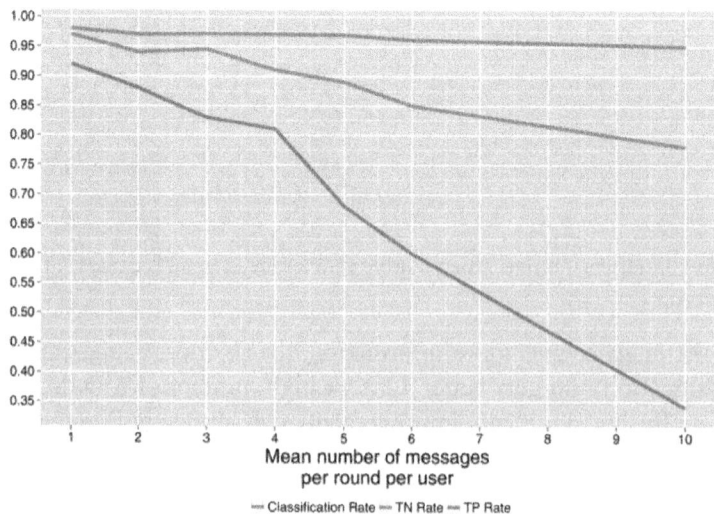

Figure 7: Classification rates vs. the mean number of messages per round.

The complexity of the problem is also related to the number of users, as can be seen in Figure 8, where the classification rate decreases as the number of users increases.

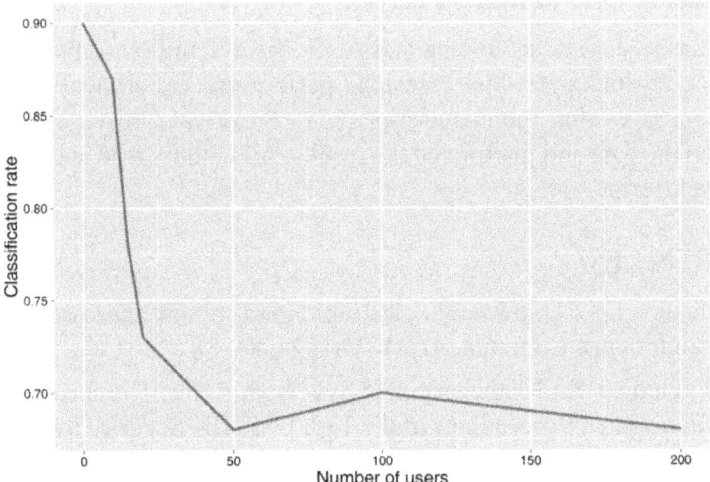

Figure 8: Classification rate vs. the number of users.

CONCLUSIONS

This work presents a method to detect relationships (or non-existent relationships) between users in a communication framework, when the retrieved information is incomplete. The method can be extended to other settings, such as pool mixes, or situations where additional information can be used. Parallel computing has also been successfully used in order to obtain faster results. The method can also be used for other communication frameworks, such as social networks or peer-to-peer protocols, and for real de-anonymization problems not belonging to the communications domain, such as disclosing public statistical tables or forensic research. More research has to be done involving the selection of optimal cut points, the optimal number of generated tables or further refinements of the final solution, which may be through the iterative filling of cells and cycling the algorithm.

ACKNOWLEDGMENTS

Part of the computations of this work were performed in EOLO, the HPC of Climate Change of the International Campus of Excellence of Moncloa, funded by MECD and MICINN. This is a contribution to CEI Moncloa. The authors would like to acknowledge funding support provided by Red Garden

Technologies Mexico. Also, the authors thank the comments of MSc. Facundo Armenta Armenta for his valuable feedback to carry out this project.

AUTHOR CONTRIBUTIONS

J. Portela, L. J. García Villalba and A. G. Silva Trujillo are the authors who mainly contributed to this research, performing experiments, analysis of the data and writing the manuscript. A. L. Sandoval Orozco and T.-H. Kim analyzed the data and interpreted the results. All authors read and approved the final manuscript.

REFERENCES

1. Chaum, D.L. Untraceable electronic mail, return addresses, and digital pseudonyms. Commun. ACM 1981, 24, 84–88.

2. Dingledine, R.; Mathewson, N.; Syverson, P. Tor: The second generation onion router. Proceedings of the 13th USENIX Security Syposium, 9–13 August 2004; pp. 303–320.

3. Pfitzmann, A.; Pfitzmann, B.; Waidner, M. ISDN-Mixes : Untraceable communication with small bandwidth overhead. In GI/ITG Conference: Communication in Distributed Systems; Springer-Verlag: Berlin, Germany, 1991; pp. 451–463.

4. Gulcu, C.; Tsudik, G. Mixing E-mail with Babel. Proceedings of the Symposium on Network and Distributed System Security, San Diego, CA, USA, 22–23 February 1996; pp. 2–16.

5. Moller, U.; Cottrell, L.; Palfrader, P.; Sassaman, L. Mixmaster Protocol Version 2. Internet Draft Draft-Sassaman-Mixmaster-03, Internet Engineering Task Force, 2005. Available online: http://tools.ietf.org/html/draft-sassaman-mixmaster-03 (accessed on 9 February 2015).

6. Danezis, G.; Dingledine, R.; Mathewson, N. Mixminion: Design of a type III anonymous remailer protocol. Proceedings of the 2003 Symposium on Security and Privacy, Oakland, CA, USA, 11–14 May 2003; pp. 2–5.

7. Serjantov, A.; Sewell, P. Passive Attack Analysis for Connection-Based Anonymity Systems. Proceedings of European Symposium on Research in Computer Security, Gjovik, Norway, 13–15 October 2003; pp. 116–131.

8. Raymond, J.F. Traffic analysis: Protocols, attacks, design issues, and open problems. Proceedings of the International Workshop on Designing Privacy Enhancing Technologies: Design Issues in Anonymity and Unobservability, Berkeley, CA, USA, 25–26 July 2000; pp. 10–29.

9. Agrawal, D.; Kesdogan, D. Measuring anonymity: The disclosure attack. IEEE Secur. Priv. 2003, 1, 27–34.

10. Danezis, G. Statistical Disclosure Attacks: Traffic Confirmation in Open Environments. InProceedings of Security and Privacy in the Age of Uncertainty; De Capitani di Vimercati, S., Samarati, P., Katsikas, S., Eds.; IFIP TC11, Kluwer: Athens, Greece, 2003; pp. 421–426.

11. Danezis, G.; Serjantov, A. Statistical disclosure or intersection attacks on anonymity systems. Proceedings of the 6th International Conference on Information Hiding, Toronto, ON, Canada, 23–25 May 2004; pp. 293–308.

12. Mathewson, N.; Dingledine, R. Practical Traffic Analysis: Extending and Resisting Statistical Disclosure. Proceedings of Privacy Enhancing Technologies Workshop, Toronto, ON, Canada, 26–28 May 2004; pp. 17–34.

13. Danezis, G.; Diaz, C.; Troncoso, C. Two-sided statistical disclosure attack. Proceedings of the 7th International Conference on Privacy Enhancing Technologies, Ottawa, ON, Canada, 20–22 June 2007; pp. 30–44.

14. Troncoso, C.; Gierlichs, B.; Preneel, B.; Verbauwhede, I. Perfect Matching Disclosure Attacks. Proceedings of the 8th International Symposium on Privacy Enhancing Technologies, Leuven, Belgium, 23–25 July 2008; pp. 2–23.

15. Danezis, G.; Troncoso, C. Vida: How to Use Bayesian Inference to De-anonymize Persistent Communications. Proceedings of the 9th International Symposium on Privacy Enhancing Technologies, Seattle, WA, USA, 5–7 August 2009; pp. 56–72.

16. Kesdogan, D.; Pimenidis, L. The hitting set attack on anonymity protocols. Proceedings of the 6th International Conference on Information Hiding, Toronto, ON, Canada, 23–25 May 2004; pp. 326–339.

17. Mallesh, N.; Wright, M. The reverse statistical disclosure attack. Proceedings of the 12th International Conference on Information Hiding, Calgary, AB, Canada, 28–30 June 2010; pp. 221–234.

18. Bagai, R.; Lu, H.; Tang, B. On the Sender Cover Traffic Countermeasure against an Improved Statistical Disclosure Attack. Proceedings of the IEEE/IFIP 8th International Conference on Embedded and Ubiquitous Computing, Hong Kong, China, 11–13 December 2010; pp. 555–560.

19. Perez-Gonzalez, F.; Troncoso, C.; Oya, S. A Least Squares Approach to the Static Traffic Analysis of High-Latency Anonymous Communication Systems. IEEE Trans. Inf. Forensics Secur. 2014, 9, 1341–1355.

20. Oya, S.; Troncoso, C.; Pérez-González, F. Do Dummies Pay Off? Limits of Dummy Traffic Protection in Anonymous Communications. Available online: http://link.springer.com/ chapter/10.1007/978-3-319-08506-7_11 (accessed on 3 February 2015).

21. Mallesh, N.; Wright, M. An analysis of the statistical disclosure attack and receiver-bound.Comput. Secur. 2011, 30, 597–612.

22. Chen, Y.; Diaconis, P.; Holmes, S.P.; Liu, J.S. Sequential Monte Carlo methods for statistical analysis of tables. J. Am. Stat. Assoc. 2005, 100, 109–120.

23. Rapallo, F. Algebraic Markov Bases and MCMC for Two-Way Contingency Tables. Scand. J. Stat.2003, 30, 385–397.

24. Greenhilla, C.; McKayb, B.D. Asymptotic enumeration of sparse nonnegative integer matrices with specified row and column sums. Adv. Appl. Math. 2008, 41, 459–481.

Chapter 11

BIDIRECTIONAL COMMUNICATION SYSTEM ON POWER LINE INTEGRATED ON ELECTRONIC BOARD FOR DRIVING OF LED AND HID LAMPS

P. Visconti,[1] S. D'Amico,[1] A. Baschirotto,[2] D. Romanello,[3]
P. Costantini,[1] V. Ventura,[1] and G. Cavalera[3]

[1]Department of Innovation Engineering, University of Salento, 73100 Lecce, Italy

[2]Department of Physics, University Bicocca, 20126 Milano, Italy

[3]Electronic Division, Cavalera Sistemi s.r.l., Galatone, 73044 Lecce, Italy

ABSTRACT

We present the bidirectional power line communication system developed in parallel to an electronic board for driving and control of HID (high-intensity discharge) and LED (light-emitting diode) lamps. The communication system, developed to be applied in the sector of public illumination, is been designed to combine high efficiency and reliability with low production costs; it consists indeed of discrete cheap components. The communication system described in this paper implements the technique of transporting digital information over existing power lines, avoiding the issue of installing new cables. Digitized signals can use power line cables through the amplitude voltage and current modulation. The solution proposed is more advantageous compared to communication techniques currently on the market which are essentially two types, power line carrier (modem for high-voltage lines) or radio (zig-Bee transceiver).

INTRODUCTION

Public lighting represents a primary service in the management of a city. It is indeed public safety guarantor, helps to improve the environment in which people live, and promotes city image through the enhancement of cultural heritage. Geographical extension of public lighting system makes it very difficult in order to ensure proper levels of efficiency, quality, reliability, and energy savings of the service by using traditional instruments. An intelligent electronic system for the control of both LED and HPS lamps for public lighting was designed. The feature of driving both types of lamps is unique for the realized system. This versatility is achieved through the implementation of a reconfigurable output capable of driving, by the user's choice, a LED or HPS lamp. Through an appropriate routine implemented in the firmware of the microcontroller, the system is capable of self-learning the type of lamp inserted in reconfigurable output. This feature allows to use the ballast in lighting systems that nowadays use the traditional discharge lamps and, primarily, will allow to keep the same ballast when discharge lamps will be replaced by the LED modules, in the near future in which the LED street lighting systems will be more affordable.

The lighting implants are indeed characterized by a large number of hot spots scattered throughout the territory for which it is constantly necessary to monitor the operation and make its management. Particularly in urban areas, public lighting systems are greatly extended and widely distributed, and they contain a large number of control panels and, above all, light points. Therefore, it is extremely problematic and expensive to make early detection of malfunctioning components, to locate and remove the cause of failures, to carefully manage the lighting implant, and to control the luminous flux of each unit. All this involves high costs in terms of time and money.

The remote control is a system designed for managing lighting systems and helps to combine cost savings with safety and service continuity. The main advantage is the implant's control in real time plus conditions improving of maintenance (i.e., real-time report of malfunctions). The remote control allows(i)eliminating unnecessary costs due to the troubleshooting,(ii) transparency, reduced operating costs, and warehouse management,(iii) automatic service regulation in the seasonal demand,(iv)real-time report on plants disruption and therefore the possibility of quick intervention to restore normal conditions,(v)scheduling of major repairs,(vi)human resources and emergency response teams optimization.

The designed transceiver system was thought to allow bidirectional communication from the control panel (master) to the individual light points (slave), using the power line as transmission medium in order to get remote control and regulation of the light intensity produced by each lighting point, as required by actual regulations regarding light pollution and energy saving.

The designed transmission system, applied to a public lighting system, allows transmission of information between the control panel and the lighting points, through the power line; consequently it is not required the installation of additional cables. This peculiarity, together with the use of relatively economical discrete components for the realization of the transmitter and receiver units, ensures low production costs, simple construction and maintenance of the system and a good overall reliability; this is certainly the main advantage of the designed communication module.

ELECTRONIC BOARD DESIGNED FOR DRIVING HID AND LED LAMPS

This section provides a brief description of the designed electronic ballast. Figure 1 shows the complete block diagram of the electronic ballast for HID and LED lamps, with the communication module highlighted. This electronic system is designed to power-supply a discharge lamp or LED lamp. The developed communication system implements the master-slave mechanism; therefore, the transceiver inserted into the ballast is the slave module. The power factor corrector (PFC) optimizes power consumption by the power line, avoiding unnecessary loss of power and provides a stabilized high voltage (Vbus). The output stage, consisting of an H-bridge, is driven by two half-bridge drivers and allows the lamp to be powered by an AC and low-frequency signal. The drivers are driven by a controller circuit which has the task of controlling the current in the lamp.

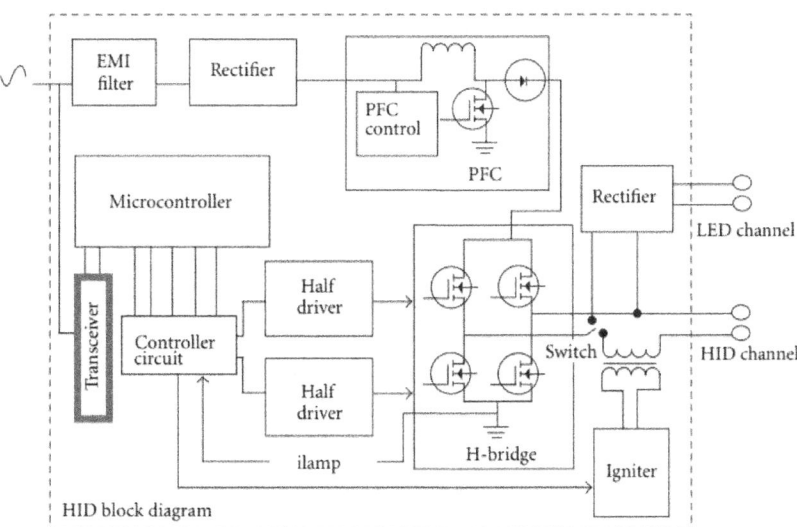

Figure 1: Electronic board block diagram.

The microcontroller provides the PWM signals to the controller circuit. The microcontroller, installed on the ballast, is intended only to generate square-wave signals to drive the H-bridge and the PWM signal in order to adjust the current in the lamp, according to the commands received from the electrical cabinet through the transceiver module. The heart of the ballast is the controller circuit: it is composed of operational amplifiers, comparators, and a programmable device (GAL). The GAL implements a logic circuit which has the aim of adjusting the current flowing in the HID/LED lamp, on the basis of signals provided by the microcontroller and the feedback from the powered lamp. The fact that this feedback does not pass through the microcontroller guarantees a certain robustness of the system. The attractive aspect of the designed board is the possibility to drive both HID and LED lamps with the same driver output. This is possible using a switch to remove temporarily the igniter circuit for LED lamp. On the LED channel is present a rectifier bridge in order to rectify the AC voltage from the H-bridge. Another important feature is the phase of autolearning of the type of load placed on the reconfigurable output of the ballast. Supplying the load with a voltage higher than the $V_{forward}$ of the eventual inserted LED module, the system begins to read the current flowing on the same load; if a current value is detected, an LED lamp is connected, and if no current is detected, it means that an HID lamp has been inserted or the channel is empty (i.e., when the HID lamp is not triggered it is an open circuit). At this point, if the load is an LED module, the process performs the soft start phase, in which a ramp current on the LED load is provided until the rated current of the LED module is reached. Once this phase is over, the load current is kept constant (burn phase). If, instead, the inserted load is recognized as an HID lamp, the first phase to perform is to create the ignition pulse of 4-5 kV across the lamp. If the lamp ignites, a warm-up phase is performed with the purpose to keep the current constant, while the voltage reaches the lamp nominal value. After this stage, the power dissipated by the lamp is kept constant (burn phase).

The designed electronic system has been tested by triggering and supplying a 150 W discharge lamp (Figure 2). By means of a DIP switch connected to the microcontroller it was possible to change the current in the lamp. In according to the state of the DIP switch, in fact, the microcontroller varies the duty cycle of the PWM signal which has the task of regulating the current in the lamp. As expected, a variation of the luminous flux produced by the lamp was observed by changing the status of the switches.

For driving with the same driver of both HID and LED lamps, the principle idea is using the H-bridge for supplying LED lamps. This type of lamps is completely different from HID lamps, and they do not require ignition circuit (disabled for this application). In Figure 3, we show an example of a LED module driven by modified H-bridge through diode-bridge rectifier.

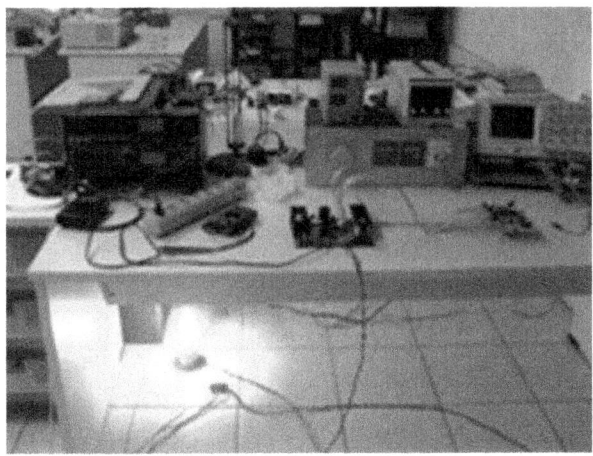

Figure 2: Experimental test.

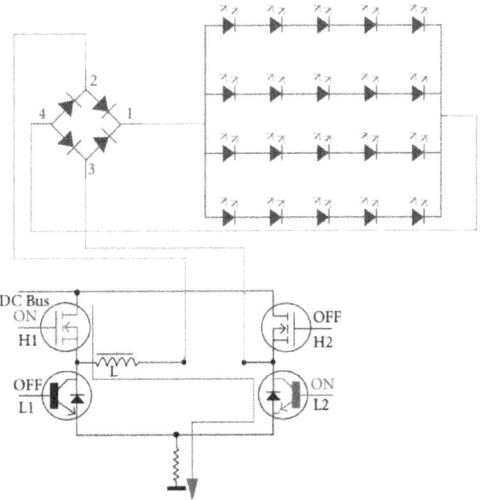

Figure 3: Led module driver circuit.

The designed electronic ballast, in case of power supplying of LED module, is not very efficient. Its main application is in fact relative to the power supply of HID lamps; in this case the system's efficiency is above 90%, as discussed elsewhere [1–3].

We have also designed an application-specific integrated circuit (ASIC), which integrates some devices installed on the ballast for HID and LED lamps, for minimizing the cost of the board and PCB area. It is fabricated using AMS

CMOS 0.35-μm technology. Then it was designed another driving board with the fully integrated ASIC that incorporates the controller circuit to adjust the current in the lamp and the control circuit of the PFC (Figure 4).

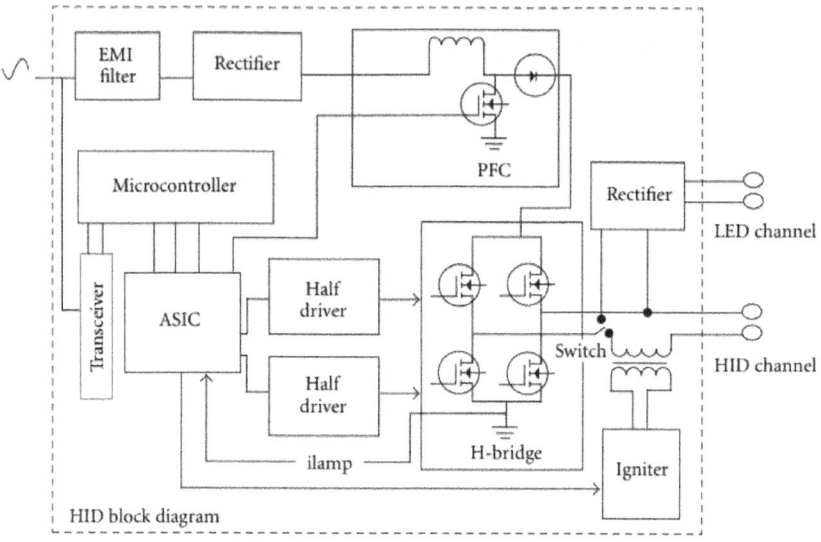

Figure 4: Electronic board block diagram with ASIC solution.

The microcontroller, through the communication module, receives the command from the electrical cabinet for adjustment of luminous flux of the lamp. Communication between the microcontroller and the ASIC is through the SPI serial communication protocol. The microcontroller provides to the ASIC the appropriate signals for the current adjustment in the lamp.

"POINT-TO-POINT" LIGHT CONTROL SYSTEM

This remote control system consists of electronic equipments for monitoring, management, and control of individual lamps; it is based on power line carrier technology that enables digital communication between the module installed on a single light point (located in the cockpit, in the terminal, or in the luminaire itself) and the managing server, placed within the command framework (Figure 5).

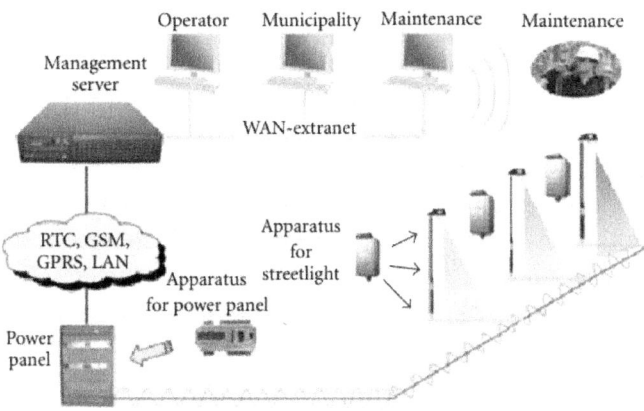

Figure 5: Light point to light point control module.

In actual power line communication system, the digital data are modulated (frequency modulation) on power supply; so there is no need of additional pipes/cables in the implant. By means of this system, it is possible monitoring and control of the electrical parameters of the individual lamps, identifying faults and alarms, turning on, turning off, minimizing the consumption, or adjusting the intensity of individual lamps using manual or automatic controls. Single light-point electrical information is transmitted periodically and stored in the control panel. By using the management software, the control system acquires the measurements and generates events and alarms based on customizable criteria. It's possible therefore managing from a central unit the single lighting point, providing an intelligent control of lighting (Figure 6).

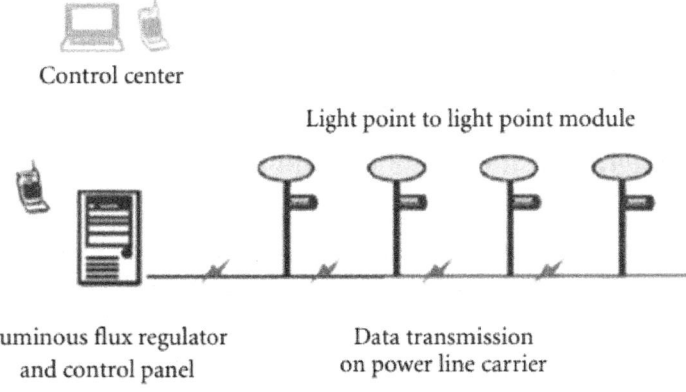

Figure 6: Light point to light point functional diagram.

COMMUNICATION SYSTEM: DESIGN CONCEPT

In this section, we present the work on the communication system between the control panel and the electronic ballast for HID and LED lamps. Designed solution is a modification of the M-Bus system characterized by the transmission between master and slave on two wires, so that communication between control panel (master) and the electronic ballast of each lamp (slave) can be realized on the power line (Figure 7). As explained above, the simplicity and low cost of this solution, in addition to good efficiency and reliability, makes it innovative compared to communication techniques currently on the market.

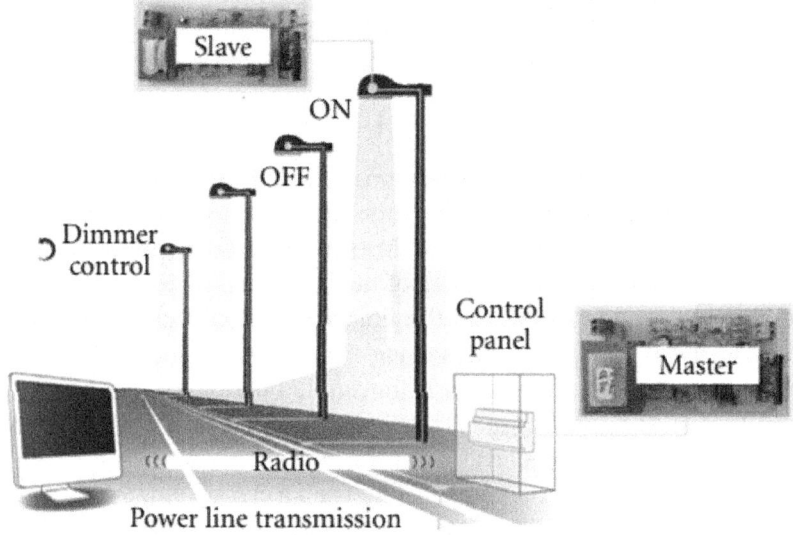

Figure 7: The communication system that we have designed, applied to a lighting system: the master will be placed in control panel, and the slave will be installed in the individual light points.

Block Diagram of the Communication System

The block diagram of the communication system is shown in Figure 8. The communication from master to slave is achieved by amplitude modulation of the voltage signal (by means of amplitude signal voltage modulatorblock); the communication from slave to master is obtained by modulating the current signal with amplitude signal current modulator block.

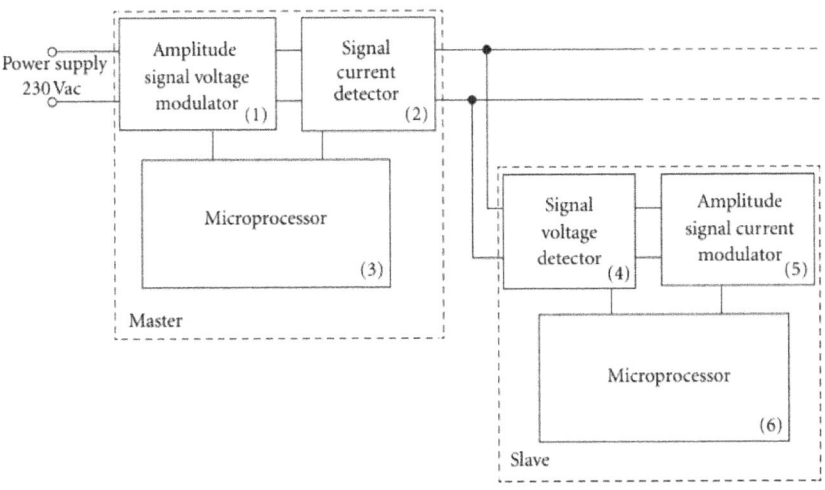

Figure 8: Transceiver block diagram.

MASTER-TO-SLAVE COMMUNICATION

The master-to-slave communication is realized through the voltage modulation on the power line.

Master-to-Slave Communication: Block Diagram

The transmission of logical 0 or 1 is generated by a microcontroller installed on the master, properly programmed to produce a digital signal that can control a switching system, that handles the insertion of a dummy load on power line, in series with the lamp. The connection of the dummy load on the line causes a voltage drop and consequently a lower network voltage on the load lamp. The network signal consists of a sine wave voltage with 230 Vrms and 50 Hz; so the transmitter PIC's digital signal is synchronized with the wave period, in order to modulate the peak values. In this way, when the dummy load is connected, we obtain reduced peak values, corresponding to transmission of a logical zero; instead, when the dummy load is not connected, we obtain unchanged peak values, corresponding to transmission of a logical one. The reception is performed by a second PIC installed on the slave, after reduction and rectification of the signal from the power line (Figure 9). The receiving PIC, synchronized by an appropriate interrupt signal, is programmed to receive this signal, to acquire the peak values and to convert them into digital format for software processing. The intelligent device, using an algorithm dedicated to this function, is thus able to discriminate the transmitted logic level.

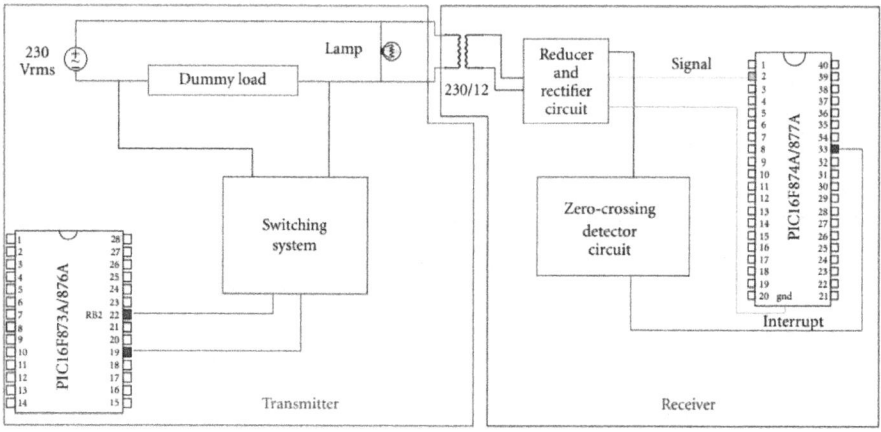

Figure 9: Simplified schematic of the master-to-slave communication system.

The slave-to-master communication is achieved by modulating the current on the power line, by arming and disarming of an additional load in parallel with the lamp of each light point. When the load is engaged, the power line current is greater than when the same load is not connected. On each light source is installed a microcontroller that, suitably programmed, manages the placement of the load according to the bits to be transmitted. In order to transmit a logical one, it inserts the load by activating the switch, through the dedicated output pins; instead to transmit a logical zero, the output pin does not allow the activation of the switch.

Master-to-Slave Communication: Algorithm Operation

Each data packet transmitted from the modulator begins with a sequence of bits called clock run-in (CRI). This sequence is designed to "announce" the start of the message and to inform the demodulator about the actual voltage level corresponding to the logical zero and logical one (depending on variation of 50 Hz network power voltage caused by different loads). The sequence of bits of the CRI in transmission is 1010101010101010. In order to carry out the experimental tests, in the packet to send, immediately after the CRI sequence, it was added, for example, the message 00011000 (Figure 10). The receiver will recognize the message transmitted according to the voltage levels provided by the CRI. The microcontroller installed in the modulator, therefore, has been programmed with a firmware capable of reproducing the signal of Figure 11.

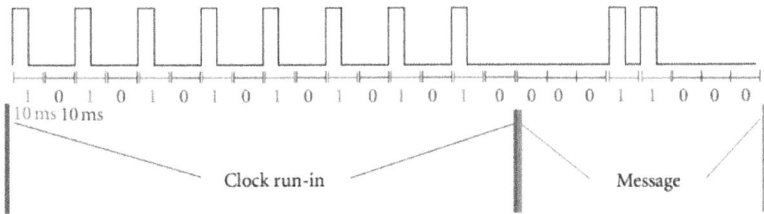

Figure 10: Signal sent from the transmitter to the receiver (clock run-in + message).

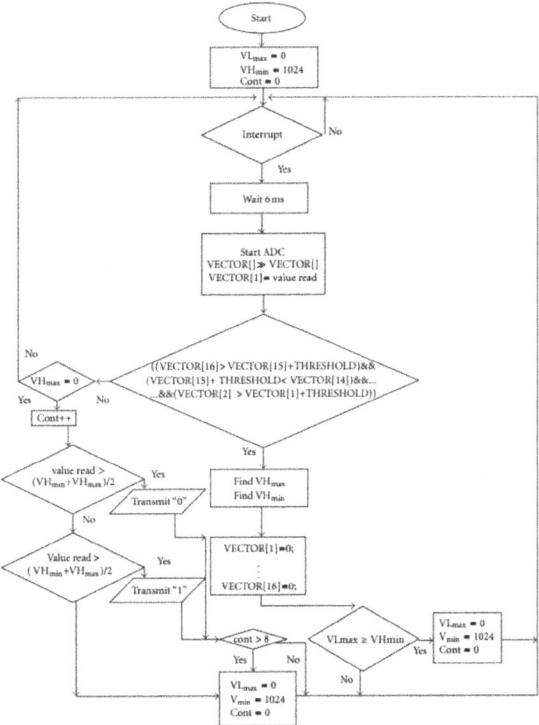

Figure 11: Algorithm of the receiver firmware.

The algorithm implemented in the firmware of the receiver is shown in Figure 11.

The receiver captures the peak voltage every 10 ms. The acquired samples are inserted into a vector (VECTOR []) in order to recognize the clock run-in.

The receiver, once recognized the clock run-in, must subsequently interpret the logic levels, 0 and 1, that characterize the message transmitted. Between the elements of vector containing the CRI are detected V_{Hmin} (minimum level of quantization in order to recognize the sample captured as logical 1) and V_{Lmax} (maximum level of quantization in order to recognize the sample captured as a logical 0). We defined a discriminating cutoff between the set of samples corresponding to the logical zero and those corresponding to a logical one, the half sum between V_{Lmax} and V_{Hmin}. Essentially, if the value acquired after the CRI is greater than or equal to $(V_{Lmax} + V_{Hmin})/2$, it is decoded as a logical one; if it is lower, consequently it is decoded as a logical zero.

Master-to-Slave Communication: Experimental Test

We have realized on breadboard the transmitter circuit consisting of the dummy load and the switching system. In Figure 12 is shown an image of the obtained electronic board.

Figure 12: Modulator circuit voltage.

In Figure 13 is illustrated the experimental set up of the communication test from master to slave.

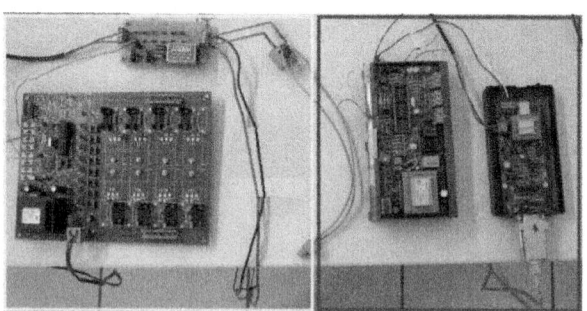

Figure 13: The workbench with the designed communication system.

In the yellow box, there are the components of the transmitter side, in particular the modulator circuit just described. The electronic board below in the yellow box, realized by Cavalera Sistemi for other applications, has been used only to the benefit of the PIC installed on it. The output voltage from the modulator (master) is sent to an incandescent lamp that represents the load on the power line. In the blue box, there is the circuitry concerning the receiver (slave). In addition to the CS083 electronic board, which contains the PIC and the zero-crossing detector circuit, there is also a RS485-RS232 converter for serial communication with the PC.

At this point, we tested the master-to-slave communication system by sending different massages and checking the received ones to the slave receiver. It was revealed that 100% of the transmitted messages has been correctly decoded (sent message 00011000 after each clock run-in). Afterwards, it was decided to verify the behavior of the communication system dynamically changing the message to be transmitted. For this scope, it was modified the firmware in the transmitter in order to send a cyclical count from 0 to 255. The test result was positive (100% of the sent messages correctly received).

Subsequently, we proceeded to test the robustness of the designed system with respect to external disturbances, such as variations of network power voltage in the range of 20% in negative and positive, simulated in experimental test by the action of the variac control, or the presence of a source of noise, generated in the test with an electric trepan connected to the same communication line. In Figure 14 is reported the shapes of transmitted digital signal and of the modulated signal. The difference between the peak values of half waves with modulation (logical 0) and without modulation (logical 1) is about 35–40 mV, over 20 quantization intervals for internal PIC 10 bit ADC.

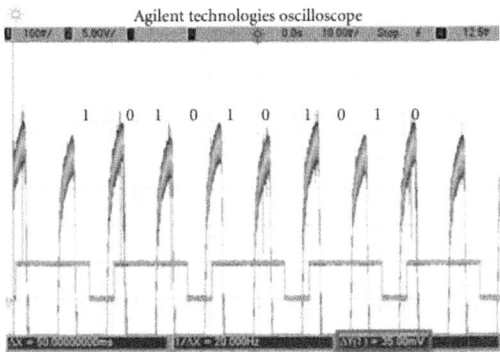

Figure 14: The signal generated by the transmitter (green) and the modulated voltage signal (yellow).

SLAVE-TO-MASTER COMMUNICATION

The slave-to-master communication is obtained by modulating the current on the power line, by arming and disarming of an additional load in parallel with the lamp (Figure 15).

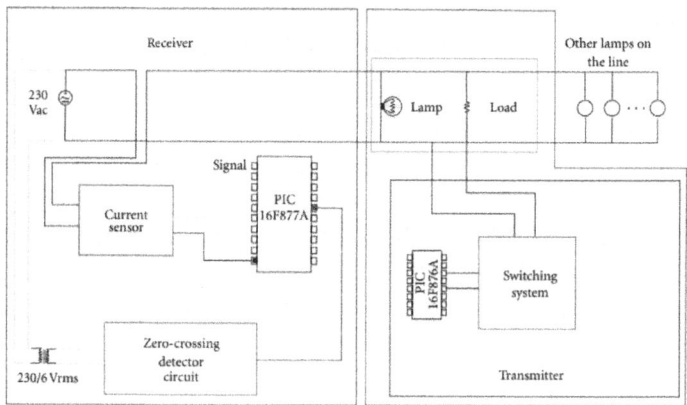

Figure 15: Simplified schematic of the slave-to-master communication system.

Slave-to-Master Communication: Block Diagram

When the load is engaged, the power line current is greater than when the same load is not connected. The produced current variation is read by the control station through a bidirectional current sensor, which produces an output voltage signal proportional to the measured current. The input-output characteristic of the current sensor is as follows. For zero current, the output is equal to 2.5 V. For positive currents the output changes to $(2.5\,V - \Delta V)$. Instead for negative currents, the output changes to $(2.5\,V + \Delta V)$. Therefore, the sinusoidal input current signal produces a sinusoidal output voltage signal, centered on the value of 2.5 V. This signal is then processed by the receiver microcontroller, that, thanks to a proper timing by the interrupt signal, captures the peak values and converts them into digital format for software processing. The PIC, properly programmed, is able to discriminate the transmitted logic level. In the slave-to-master transmission, the object of modulation, that is, the current, varies depending on the number of powered lamps; if this number increases, the current that has to be modulated increases too. In order to manage this dynamic situation, maintaining a good resolution in the transmission, it is necessary only to change the reference voltages of the AD-Converter of the microcontroller for different lighting plants.

Slave-to-Master Communication: Algorithm Operation

Regarding the algorithm of the transmitter in order to carry out the transmission tests in the laboratory, the PIC is programmed in order to generate the message shown in Figure 11. Due to the nature of the voltage signal provided by the current sensor, it was appropriate in this case to modulate the signal during the whole period, that is, every 20 ms. This of course implies a lower transmission rate. Regarding the receiver firmware, however, the only variation from to the algorithm of Figure 12 is due to an additional control; this modification was necessary because the PIC does not recognize all the peaks of the half waves but only the superior peaks.

Slave-to-Master Communication: Experimental Test

An experimental test was carried out for the communication from slave-to-master. In order to get 1 A of current in the communication line (as equivalent to one lighting point turned on in a public lighting system), we used a load consisting of four 230 V/60 W lamps connected in parallel. The slave-to-master communication system on the test bench is shown in Figure 16.

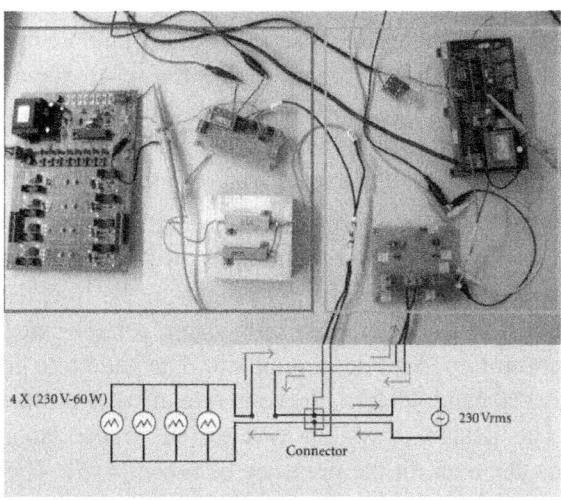

Figure 16: The "slave-to-master" communication system.

The transmitter side (contained in the lighting point) was evidenced in the figure by a red border, while the receiver side (contained in the control cabinet) with a green border. The red border shows the solution used for the resistive load that is connected/disconnected to modulate the current; it consists of two resistors in parallel, which realize an equivalent resistance of 500 Ohm–100 W. The resistors, fixed on a heat sink, are connected to the power line through the switching circuit. When they are connected to the power line, the additional resistive load will cause a current increase of 0.5 Amp (from 1 Amp to 1.5 Amp), and the sensor current will detect this variation.

The transmitter has been programmed to send the binary digits of 8 bits, for a cycle count from 0 to 255. It was observed that the transmission was successful (100% of the sent messages correctly received). The test was repeated simulating the large swings of the voltage on power line; even in these cases we have obtained satisfactory results.

CONCLUSION

We have designed and developed a communication system, on the power line, capable of allowing the bidirectional communication between the control panel (master) of a lighting plant and the single lighting point (slave). Slave-to-master communication is implemented modulating the voltage amplitude, while the slave-to-master communication is done modulating the current amplitude. This communication system has been tested in the laboratory, in both directions, connecting through cable lines long about two meters, slave circuits with the master electronic board. We have carried out several tests with different loads on the power line, in order to simulate the presence of one or more light sources (slave). Regarding the master-to-slave communication, being the lighting points connected in parallel and the communication carried out by means of amplitude modulation of the voltage, there are no limitations for the number of slaves that the system can support.

Instead, in the slave-to-master transmission, since the modulation occurs on the current, we must consider the number of lighting points on the line. In fact, according to the number of light points on the line, different solutions have been studied in order to keep the system reliable; at the moment, we achieved the control up to 40 lighting points. The baud rate in Master to Slave transmission is 100 baud; instead, the baud rate in slave-to-master transmission is 50 baud. The designed system was tested in the laboratory obtaining satisfactory results, even in the presence of noise [4, 5]. The next step is the test of the communication system on a real lighting system to determine the technical limits (transmission distance, number of slave modules supported, baud rate error, etc.) of the designed communication system.

ACKNOWLEDGMENTS

Intervention was cofinanced by the E.U.-E.R.D.F. on the O.P. Apulia Region 2007–2013-Axis I-Line 1.1 "Aid for investment in research for SMEs" and supported by Cavalera Sistemi Srl-Galatone (LE).

REFERENCES

1.   P. Visconti, D. Romanello, G. Zizzari, and G. Cavalera, "Electronic board for driving of HID and LED lamps with auxiliary power supply from solar panel and presence detector," in Proceedings of the 10th International

Conference on Environment and Electrical Engineering (EEEIC.EU '11), pp. 430–433, Rome, Italy, May 2011. ·

2. P. Visconti, D. Romanello, G. Zizzari, and G. Cavalera, "Multichannel electronic board for control of LED and high intensity discharge lamps," in Proceedings of the International Conference on Industrial Electronics and Systems Engineering (ICIESE ‹11), Venice, Italy, April 2011.

3. Costantini, D. Romanello, P. Visconti, et al., "Electronic board for driving and control of HID and LED lamps in lighting systems," in Proceedings of the 4th International Conference on Sensing Technology (ICST ‹10), S. C. Mukhopadhyay, A. Fuchs, G. S. Gupta, and A. Lay-Ekuakille, Eds., pp. 199–204, University of Salento, Lecce, Italy, June 2010.

4. X. Carcelle, Power Line Communications in Practice, Electricité de France, 2009.

5. H. Ferreira, H. Grove, O. Hooijen, and A. J. H. Vinck, "Power line communication," in Encyclopedia of Electrical and Electronics Engineering, J. G. Webster, Ed., pp. 706–716, Wiley-Interscience, New York, NY, USA, 1999.

Chapter 12

A FRAMEWORK FOR UWB-BASED COMMUNICATION AND LOCATION TRACKING SYSTEMS FOR WIRELESS SENSOR NETWORKS

Juan Chóliz, Ángela Hernández and Antonio Valdovinos

Research Institute of Engineering in Aragón, I3A, University of Zaragoza, C/María de Luna 3, Zaragoza 50018, Spain

ABSTRACT

Ultra wideband (UWB) radio technology is nowadays one of the most promising technologies for medium-short range communications. It has a wide range of applications including Wireless Sensor Networks (WSN) with simultaneous data transmission and location tracking. The combination of location and data transmission is important in order to increase flexibility and reduce the cost and complexity of the system deployment. In this scenario, accuracy is not the only evaluation criteria, but also the amount of resources associated to the location service, as it has an impact not only on the location capacity of the system but also on the sensor data transmission capacity. Although several studies can be found in the literature addressing UWB-based localization, these studies mainly focus on distance estimation and position calculation algorithms. Practical aspects such as the design of the functional architecture, the procedure for the transmission of the associated information between the different elements of the system, and the need of tracking multiple terminals simultaneously in various application scenarios, are generally omitted. This paper provides a complete system level evaluation of a UWB-based communication and location system for Wireless Sensor Networks, including aspects such as UWB-based ranging, tracking algorithms, latency, target mobility and MAC layer design. With this purpose, a custom simulator has been developed, and results with real UWB equipment are presented too.

INTRODUCTION

The use of location and tracking information is an excellent tool to improve productivity and to optimize resource management in a wide range of sectors: industrial, medical, home-automation or military. Whereas satellite systems, i.e., GPS, are widely used in outdoor applications such as vehicle navigation, fleet management or emergency call localization, there are multiple alternatives for the development of indoor Location & Tracking (LT) systems. Despite not being specifically designed for that purpose, various widespread radio technologies such as cellular (GSM, UMTS, LTE) and short-medium range wireless systems (WiFi, Bluetooth, ZigBee, RFID), may provide location information with different levels of accuracy, range and complexity [1,2]. Within this group of short range radio systems, Ultra-Wideband (UWB) stands out providing high accuracy on distance estimation with remarkable features concerning size and power consumption and allowing simultaneous location and data transmission [3].

IR (Impulse Radio) UWB communication systems are based on the transmission of very short duration pulses, which originate very high bandwidth signals. The short duration of the pulses allows a high level of accuracy in time of arrival estimation and as a result a centimeter-level resolution in distance estimation (ranging). Furthermore, due to the short duration of the transmitted pulses, UWB provides unmatched performance on multipath and NLOS environments [4]. In addition, low complexity and low power consumption of UWB transceivers is essential in order to design battery-powered sensors.

In general, location determination comprises two phases, angle and/or distance estimation and position calculation. Angles and distances between the element to be located and some fixed reference nodes can be estimated based on the measurement of different parameters such as Angle of Arrival (AOA), Received Signal Strength Indication (RSSI) and Time of Arrival (TOA) of reference signals exchanged between them. In particular, TOA estimation requires the exchange of ranging frames between the element to be located and the reference nodes, which entails that some temporal resources must be dedicated to location and that a non-negligible latency is associated to the position update process. On the other hand, several algorithms can be used to compute the position according to the estimated distances or angles, including geometry-based (triangulation, trilateration, multidimensional scaling), least square and cost function minimization, fingerprint and Bayesian techniques (Kalman and particle filters).

Extensive research has focused on the design of distance estimation and position calculation algorithms in the last few years [5–8]. Nevertheless, in general these studies focus on algorithm optimization, and simple scenarios with a single terminal and a few previously defined reference nodes are considered. Only a few studies address practical aspects such as the design of the functional architecture, the procedure for the transmission of the associated information between the different elements of the system, and the need of tracking multiple terminals simultaneously in various application scenarios. These aspects would lead to consider the amount of resources associated to the global location service as a quality parameter. Moreover, a rigorous algorithm evaluation requires using a dynamic scenario with multiple mobile terminals and reference nodes.

On the other hand, although a few UWB-based LT systems can already be found in the market, these systems are proprietary solutions and only use UWB for distance estimation, while communication between the different elements involved in the positioning system is done using other technologies, generally wired. In contrast, UWB systems combining location and data transmission (for example in the framework of IEEE 802.15.4a standard) would increase flexibility reducing the cost and complexity of the system deployment.

IEEE 802.15.4 standard offers the fundamental lower network layers of wireless personal area networks (WPAN) focusing on low-cost, low-speed ubiquitous communication between devices with little to no underlying infrastructure [9]. It provides support for ad hoc networks capable of performing self-management and organization, aimed at Wireless Sensor Network (WSN) applications. IEEE 802.15.4a-2007 enhanced the standard specifying two additional PHYs: Chirp Spread Spectrum and Direct Sequence UWB, which enhances the standard with the accurate distance estimation capability of UWB [10]. This way, a UWB-based Wireless Sensor Network (e.g., a fire detection sensor network) could be simultaneously used to provide mobile users tracking. In this scenario, it becomes clear that accuracy is not the only evaluation criteria of LT systems, but also the amount of resources associated to the location service, as it has an impact not only on the location capacity of the system, but also on the consequent reduction of the available data rate for sensor data communication.

The combination of wireless sensor networks with UWB accurate location capabilities enables a wide variety of application scenarios. For example, in order to guarantee the safety of workers in dangerous environments (electrical substations, fires, accidents, etc.) tracking their position and monitoring at the same time the level of different parameters (electric field, carbon monoxide, radiation, etc.). Another example is sports tracking, in order to provide a complete monitoring of performance (distance travelled, average and peak speed, acceleration, etc.) and biometric information (heart rate, blood pressure, etc.). In industrial applications, the location of a certain product in the assembly line or in a warehouse could be monitored, together with its temperature, humidity, etc.

The main objective of this paper is to design a communication and location tracking system for Wireless Sensor Networks based on Ultra-Wideband technology and to provide a complete system-level evaluation. Besides distance estimation and location and tracking algorithms, system level evaluation includes aspects such as target mobility, functional architecture, distribution of the information related to the location function and latency of the position update process. These aspects are not usually considered in the existing studies, but have a great impact on the capacity of systems combining data transmission and location.

The paper is structured as follows. Section 2 summarizes the literature related to the presented study. In Section 3 the proposed communication and location tracking UWB system is described, including the network topology, the tracking functional architecture, the distribution and acquisition of location information and the tracking function implementation. In Section 4 the system performance is evaluated using a custom self-developed simulator, and the different system design alternatives are assessed. Measurements with real UWB equipment are also provided. Section 5 summarizes the main conclusions.

RELATED WORK

As it was previously mentioned, most of the previous works related to UWB location and tracking systems focus on distance estimation techniques and location and tracking algorithms, but only a few of them address practical aspects such as the transmission of the information associated to location among the different elements of the system. Some proposals of UWB-based communication and location tracking systems can be found in the literature. In [11] a UWB-based system for indoor location services is introduced, which relies on a three-tier hierarchical sensor architecture to cover a large indoor space, also defining the communication between the different tiers and the location procedure. In [12] the design of a UWB-based ad hoc network for search and rescue operations in disaster zones is presented, defining the network architecture, physical entities and a complete protocol stack, from the physical layer up to the application layer. In [13] a group of communication protocols and localization algorithms for wireless sensor networks in coal mine environments is proposed, specifically a new UWB coding method, an ALOHA-type channel access method and a message exchange protocol to collect location information. Finally, in [14] an overview of an IR UWB open prototyping platform that illustrates a fully integrated solution from physical layer up to application layer is provided. However, these works [11–14] mainly focus on the system description, including in some cases a very basic system evaluation. Based on the system proposed in [14], this paper presents a thorough analysis and evaluation of different aspects such as the functional architecture, the acquisition and distribution of the information associated to location and the effect of position update latency and target mobility.

The optimization of MAC layer design for simultaneous location and communication has also been addressed in different works. In [15] an early approach to UWB MAC layer issues for location and tracking applications is provided and various UWB system architectures, MAC schemes and network solutions are discussed, although no evaluation of the proposals is provided. In [16] a MAC layer design for IR UWB location networks is proposed in order to determine the locations of a network of stationary reference nodes and mobile nodes deployed in an ad hoc manner based on a small number of fixed anchors with known locations. Location and range information is propagated through the network of reference nodes to periodically estimate the locations of the mobile nodes, which entails a convergence time and associated throughput. On the contrary, the tracking system proposed in the present work considers that the locations of the reference nodes are known, whether they are known a priori or determined in a prior set-up phase. This way the convergence time and associated throughput are minimized, thus providing fast tracking of the mobile nodes. The MAC layer design we have considered as the basis for the proposed UWB communication and location tracking system was presented in [17]. This MAC is based on the IEEE 802.15.4 standard [9], although it deviates from the standard in a few areas such as the support to peer-to-peer communications, the usage of guaranteed time slots for data transmissions and dedicated time slots for ranging and allocation requests, the definition of a relaying functionality and the specification of ranging and localization procedures. In [18] the performance of this MAC is studied under the point of view of tracking, evaluating the time delay necessary to collect the ranging information as a function of the number of mobiles in the network. Furthermore, a few enhancements are proposed in order to minimize the exchange of packets necessary to update the ranging information. However, this study assumes the existence of physical connectivity between all the nodes and does not take into account the resource constraints of the MAC layer design, aspects that are considered in this study, which also evaluates the performance degradation in terms of accuracy due to the latency associated to the position update process.

WIRELESS SENSOR COMMUNICATION AND LOCA-TION TRACKING SYSTEM PROPOSAL

The proposed communication and location tracking UWB system aims to enable wireless data communication within a network of sensors and at the same time to track walking users in wide indoor areas with accuracy below 1 m. The system is composed of multiple UWB picocells. Each picocell is composed of mobile nodes to be tracked (targets) and fixed sensor nodes with known positions (anchors). Distances between the target and the anchor nodes are estimated through a ranging frame exchange. Estimated distances are sent to location controllers (LC), which are the functional units that execute the tracking algorithm to obtain the estimated position of the targets. The main characteristics of the network and the different options considered are detailed below.

Network Topology and PHY/MAC Structure

The application scenario is covered by multiple UWB picocells, although for simplicity a single picocell is considered. PHY and MAC layers of an open IR-UWB platform described in [14] are assumed. This platform is based on the 802.15.4a standard [10], although it is not fully compliant. Table 1 summarizes the main PHY and MAC parameters considered.

Table 1: PHY and MAC parameters

Parameter	Value
Frequency range	3.5–4.5 GHz
Symbol duration	2.88 μs
Raw bit rate	347 kbps
Slot length	160 bytes
Slot duration	3.686 ms
Maximum superframe length	53 slots
Beacon Interval	195.379 ms
Maximum Beacon Period length	12 slots
Max. Topology Management Period length	3 slots (12 subslots)
Maximum CFP length	26 slots
Number of slots for data communication	8 slots
Number of slots for ranging	12 slots
GTS request period length	6 slots (12 subslots)
Maximum CAP length	12 slots (48 subslots)

The picocell topology is mesh centralized, as shown in Figure 1. A picocell coordinator transmits beacon frames for common superframe synchronization and handles the scheduling procedures. Then, a scheduling tree is built and used to transport beacon and command frames, which are relayed from the picocell coordinator to any node in the picocell. Finally, it becomes a meshed scheduling tree by enabling the transmission out of the tree for the data, ranging and hello frames.

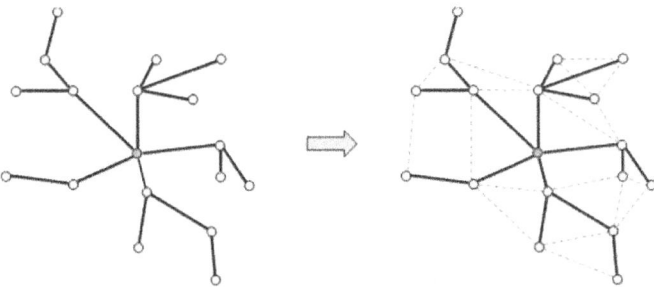

Figure 1: Mesh centralized topology.

The MAC superframe is divided into timeslots that are grouped into different periods, as it is shown in Figure 2.

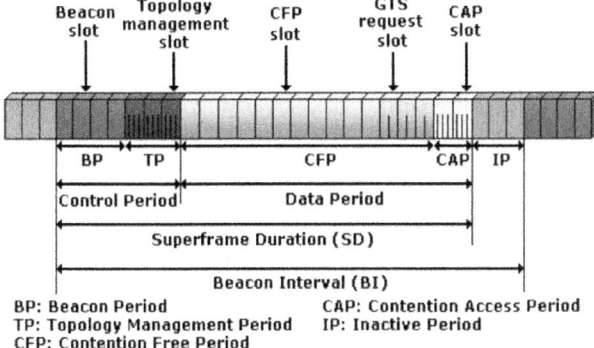

Figure 2: Proposed MAC superframe structure.

- Beacon period: Used for the beacon alignment. The first beacon slot is reserved for the coordinator.

- Topology Management Period: Used for the periodic broadcast of hello frames from each node. This way the neighborhood is known locally for each node of the network.

- Contention Free Period (CFP): It is composed of Guaranteed Time Slots (GTS) for sensor data, location data and ranging frames transmission, and a GTS request period. Concerning data frames, if source and destination nodes are not physically connected frames are relayed at MAC level using consecutive timeslots. Ranging frames are not relayed and can be sent only between neighbor nodes. Two types of ranging frames are defined: ranging request and ranging response.

- Contention Access Period (CAP): Used for the transmission of command frames through a slotted ALOHA multiple access scheme. Each CAP slot is divided into subslots in order to relay commands.

It should be noted that the relaying procedure is performed at the MAC layer level. When a node has data to transmit, it sends a GTS request on the tree to the coordinator with its address as the source address and the destination address of the transmission. The coordinator, which has the knowledge of the whole network, looks on its routing table if there are relays between the source and the destination. If there are relays, the coordinator determines the route and allocates the GTS for each link.

Functional Architecture and Strategies for Acquisition and Distribution of Location Information

In order to track the position of the target nodes, location information, basically the distances estimated between the target and the anchor nodes, must be acquired and transmitted to a LC that executes the tracking functionality. LCs can be physically located in one or more anchor nodes or in the target nodes. Depending on the location of the LC, several tracking functional architectures (centralized and distributed) can be defined. On the other hand, either the target or the anchor nodes may estimate the distance. This function is referred to as distance acquisition function. The allocation of the distance acquisition function to the target or the anchor nodes is a design alternative that may have an impact on the need of resources.

Tracking Function Distribution in the Network

Depending on the location of the LC function, different tracking functional architectures can be defined. In the tracking functional architecture that we denote as centralized architecture, the tracking functionality is implemented in one or more previously defined anchor nodes that become LCs. Figure 3 shows an example of a centralized architecture with one LC. Using one LC entails a higher need of resources, as multiple hops will be needed to forward the location information to the LC. Defining multiple LCs reduces the need of resources, but increases complexity, as the tracking functionality must be implemented in several nodes and a procedure should be implemented to assign each target to the closest LC. In the distributed architecture, each target dynamically picks one of its neighbor anchors to execute the tracking functionality. Therefore, there may be as many simultaneous LCs as targets. As the LC is always executed by an anchor neighbor to the target, only one timeslot will be needed to exchange data frames between the target and the LC, and resources will consequently be reduced. As a drawback, the tracking functionality must be implemented in every anchor.

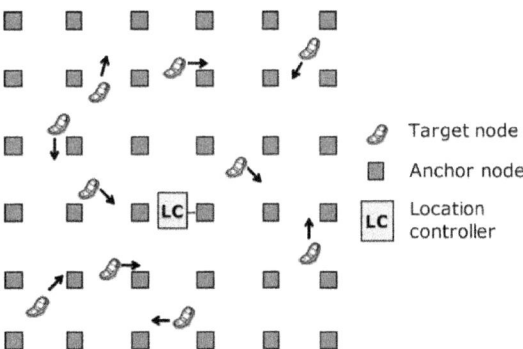

Figure 3: Tracking system. Centralized architecture with 1 LC.

Finally, in the target-centered architecture the LC function is implemented in the target nodes. The target nodes perform ranging with their neighbor anchors and obtain their own position applying the tracking algorithm. Therefore, there is no need of transmitting the estimated distances and the updated position. Nevertheless, the implementation of the tracking functionality requires certain computational capacity on the target nodes, increasing their complexity and cost.

Acquisition and Distribution of Location Information

The acquisition of the location information is done through the ranging procedure. The procedure initiator (target or anchor) transmits a ranging request to another node, which estimates the time of arrival and sends a ranging response after a predefined time. The initiator measures the time of arrival of the response and can estimate the transmission delay and the distance between the nodes (Two Way Ranging). In order to improve the accuracy of distance estimation, two ranging responses can be sent in order to compensate for the clock drift (Three Way Ranging).

Once the distances between the target and the anchor nodes have been estimated, they must be transmitted to the LC in a location data frame. The LC calculates and transmits the updated position to the target node. In order to reduce the amount of resources needed to acquire and distribute the location information, different enhancements can be applied [18].

- Data aggregation: All the distances estimated can be aggregated in a single location data frame and sent to the LC by the ranging initiator (target or anchor).

- Broadcast/multicast request: The ranging initiator (target or anchor) can aggregate multiple ranging requests into a single ranging request sent to all its neighbor nodes (broadcast request) or to a subset of them (multicast request).

- Multicast response: After receiving the ranging requests from different nodes, an anchor or target node can aggregate all the responses into a single multicast response.

If the initiator is the target node, multicast response would require the simultaneous update of the position of all the targets, so the anchors can aggregate the responses to multiple targets into a single multicast response packet. The same applies for broadcast/multicast request and data aggregation when the initiator is the anchor node.

Tracking Function Implementation

With respect to the tracking technique itself, parametric and non-parametric approaches can be distinguished. Parametric approaches compute the location based on the a priori knowledge of a model. On the other hand, non-parametric approaches do not require model knowledge, although in some cases they may use some statistic parameters (mean, variance). Specifically, the following algorithms are considered in this study: Trilateration, Least Square-Multidimensional Scaling (LS-MDS), Least Square-Distance Contraction (LS-DC), Extended Kalman Filter (EKF) and Particle Filter (PF).

Trilateration is a non-parametric algorithm that computes the position based on the distance estimated between the target and three anchor nodes using a geometrical method for determining the intersection of three sphere surfaces [7]. Consequently, regardless of the number of anchors selected, only the three anchors with smallest estimated distance to the target are used for position computation.

The algorithm LS-MDS is a completely non-parametric approach combining Multidimensional Scaling (MDS) with Least Squares (LS) minimization [19]. MDS is a multivariate data analysis technique used to map "proximities" into a space. These "proximities" can be either dissimilarities (distance-like quantities) or similarities (inversely related to distances). Given npoints and corresponding dissimilarity, MDS finds a set of points in a space such that a one-to-one mapping between the original configuration and the reconstructed one exists. Then it is possible to map back the solution to the absolute reference system by Procrustes transformation. MDS is used to obtain a previous estimation of the solution. Then, the localization problem is posed into a non linear least squares optimization problem. The goal is to obtain the matrix of computed positions X that minimizes the stress function σ (X) defined as follows:

$$\sigma(X) = \sum_{i=1}^{n} \sum_{j=1}^{n} (\partial_{ij} - d_{ij}(X))^2$$

(1)

where ∂_{ij} is the estimated distance between nodes i and j and $d_{ij}(X)$ is the distance between nodes i and j associated with the computed node locations X. In order to solve this problem, a low-complexity algorithm based on majorization technique is applied. Specifically, the algorithm is known as SMACOF and it consists of an iterative procedure that attempts to find the minimum of a non-convex function by tracking the global minima of the so-called majored convex function successively constructed from the original objective and basis on the previous solution.

LS-DC combines the Distance Contraction (DC) algorithm with Least Square minimization [20]. Firstly, the distances between the target and n anchor nodes are estimated. Each estimated distance defines a line of position for the target's location as a circle around the corresponding anchor node. Then, the feasibility region is defined as the area of intersection between the n circles. If the feasibility region does not exist, distance contraction cannot be applied and the LS-MDS approach is used instead. If the feasibility region exists, an initial point is computed inside the feasibility region and the contracted distances are computed as the shortest distance from each anchor to the aforementioned feasibility region. Finally, a minimization algorithm is applied using the contracted distances instead of the estimated ones. Since the function becomes convex, any minimization algorithm (i.e., global distance continuation, steepest descent) can be used thus reducing complexity, although here SMACOF has been used in order to be comparable with LS-MDS.

The Extended Kalman Filter is a Bayesian technique known for its low complexity, performance and stability as a tracking algorithm [21]. EKF addresses the problem of trying to estimate the state x of a discrete-time controlled process that is governed by a non-linear stochastic difference equation:

$$x_t = f\left(x_{t-1}, u_{t-1}, w_{t-1}\right)$$

(2)

with a measurement z that is:

$$z_t = h\left(x_t, e_t\right)$$

(3)

The non-linear function f in Equation (2) relates the state at the previous time step $t - 1$ to the state at the current time stept and includes as parameters any driving function u_t and the process noise w_t. The non-linear function h in Equation (3)relates the state x_t to the measurement z_t and includes as parameter the measurement noise e_t. Process and measurement noise are assumed to be independent, white, and with normal probability distributions $p(w) \in N(0, Q)$ and $p(e) \in N(0, R)$. The process and measurement noise covariance matrixes Q and R are defined by variances σ_w^2 and σ_e^2.

The Kalman-based tracking algorithm has two major stages, namely, the update and the correction stages, which are iterated k times for every observation occurring at a given time. The time update equations project the state and covariance estimates from the previous time step $t - 1$ to the current time step t. The measurement update equations correct the state and covariance estimates with the measurement z_t. As f and h cannot be applied to the covariance directly, matrixes of partial derivatives (Jacobian) are computed.

Focusing on the implementation of the EKF for the tracking application, the state vector x contains target's position p_t and speed v_t as process variables. The measure vector z contains the process observations, namely the estimated distances between the target and the anchors. The functions that describe the evolution of the state vector through time and the relation between the state vector and the measure vector are:

$$\begin{pmatrix} p_{t+1} \\ v_{t+1} \end{pmatrix} = \begin{pmatrix} I & T_s \cdot I \\ 0 & I \end{pmatrix} \begin{pmatrix} p_t \\ v_t \end{pmatrix} + \begin{pmatrix} T_s^2 / 2 \cdot I \\ T_s \cdot I \end{pmatrix} w_t$$

(4)

$$\tilde{z}_t(i) = | p_i - \tilde{p}_t | + e_t$$

(5)

where T_s is the time between two consecutive updates and p_i is the position of anchor i.

Finally, Particle Filters are recursive implementations of Monte Carlo based statistical signal processing. The use of particle filters for positioning in wireless networks was proposed in [22]. The particle filter is based on a high number of samples of the state vector or particles, which are weighted according to their importance (likelihood) in order to provide an estimation of the state vector. The advantage of particle filters over other parametric solutions is that non-linear models and non-Gaussian noise can be defined. As a drawback, their computational complexity is higher, so they are suitable in applications where computational power is rather cheap and the sampling rate slow. As for EKF, a state vector x, a measure vector z and functions f and h are defined. On each step, the particles are moved according to Equation (4) and the weights are updated according to the likelihood of the observations:

$$w_t^i = w_{t-1}^i p(z_t | x_t^i), \quad i = 1, ..., N$$

(6)

where i is the particle index, N is the number of particles and the probability p(zt|xit) is equivalent to the probability pe(zt−h(xit)) according to the distribution of the measurement error e. But here the measurement noise e is not necessarily considered Gaussian. Specifically, we have defined the measurement error model as a weighted sum of three Gaussian components for the different channel configurations (LOS/NLOS/NLOS2). The weight of each component is also a Gaussian-like function as will be later defined in Equation (9). Consequently, the filter is defined by the variance of process noise σ_w^2 and the parameters of the measurement error model (mean and variance of each component and mean and variance of each weight).

PERFORMANCE EVALUATION

System Model and Simulator Description

In order to evaluate the impact of the different system design alternatives and parameters, we have developed a specific simulation application using C++. The simulation scenario represents a relatively wide indoors area, such as a warehouse, where people and goods moving at pedestrian speeds will be tracked with accuracy below 1 m. On this scenario a UWB-based wireless sensor network composed of N_a anchor nodes and N_m target nodes is deployed. The existence of walls and obstacles is considered through the use of an indoor ranging model which accounts for the probability of non-line-of-sight between targets and anchors. The dynamics of the targets are modeled by a Random Walk Model, with random directions and speeds that are constant during a certain period of time, after which new random directions and speeds are set [23].

A common set of parameters has been defined. Area size has been set to 50 m × 50 m in order to represent a relatively wide indoors area. UWB nodes range has been set to 15 m according to the specifications of the IR-UWB platform presented in Section 3.1 [14]. In order to guarantee connectivity between adjacent anchors, the distance between adjacent anchors has been set to 10 m, which results into 36 anchors. Concerning the dynamics of the target nodes, the Random Walk Model is defined by the minimum speed, the maximum speed and the change interval, that have been set to 0.1 m/s, 3 m/s and 20 s respectively in order to model pedestrian motion. Finally, as a result of prior simulations, the nominal position update interval has been set to 976 ms (5 superframes), as it provides accurate tracking of the moving targets with a reasonable use of resources (timeslots).

A ranging model is used to characterize the ranging error distribution and to generate the distance estimation samples. Range measurements based on round-trip TOA estimation through n-Way Ranging transactions can be modeled as:

$$\tilde{d}_{ij} = d_{ij} + \varepsilon_{ij} + n_{ij} = d'_{ij} + n_{ij}$$

$$(7)$$

where d_{ij} is the actual distance between nodes i and j, d_{ij}' is the biased distance (with bias ε_{ij}) and n_{ij} is a residual error term.

The biased distance is modeled as a weighted sum of Gaussian and Exponential components conditioned upon the actual distance and channel configuration. The pdf of d', conditioned upon d and a particular channel configuration C, is described as follows:

$$p_C\left[d'/(d,C)\right] = G_C \frac{1}{d}\frac{1}{\sqrt{2\pi}\sigma_C}e^{-\frac{\left(\frac{d'}{d}-1\right)^2}{2\sigma_C^2}} + E_C \frac{1_{\{d'>d\}}}{d}\lambda_C e^{-\lambda_C\left(\frac{d'}{d}-1\right)}$$

$$(8)$$

where $d \neq 0$, $\{G_C, \sigma_C\}$ and $\{E_C, \lambda_C\}$ are the weights and parameters of Gaussian and Exponential mixture components, $1_{\{x>y\}} = 1$ whenever $x > y$ and 0 otherwise, and C takes its value among {LOS, NLOS, NLOS2}. The model is enhanced by taking into account the probability $W_C(d)$ to have a particular channel configuration at a distance d. These weights are described as Gaussian-like functions:

$$W_C(d) = \frac{\xi}{\sqrt{2\pi}\varsigma_C}e^{-\frac{(d-d_C)^2}{2\varsigma_C^2}}$$

$$(9)$$

This model for the ranging bias was proposed and validated through a measurement campaign with real UWB equipment in an office environment in [24], where the values for the different parameters of the model were also identified.

The residual error is modeled as additive and centered, with a variance σ_n^2 that depends on detection error terms affecting unitary TOA estimates, i.e., receiver sampling rate, and involved protocol durations, and is independent of the distance between the nodes.

In order to set a realistic value for ranging residual error σ_n, a measurement campaign was carried out using the open IR-UWB platforms mentioned in Section 3.1. Line-of-sight and distances up to 5 m in steps of 0.25 m were considered. Under these conditions, the error is mostly due to residual error. For each distance, 150 ranging samples were obtained and the mean and standard deviation of the distance estimation error was computed. As it can be observed in Figure 4, the mean is always close to zero and no dependency on distance has been detected, as it was expected for ranging residual error. On the other hand, standard deviation varied from 10 cm to 40 cm, with an average value of 25 cm. As a result, we will consider a value of $\sigma_n = 0.3$ m for the subsequent simulations.

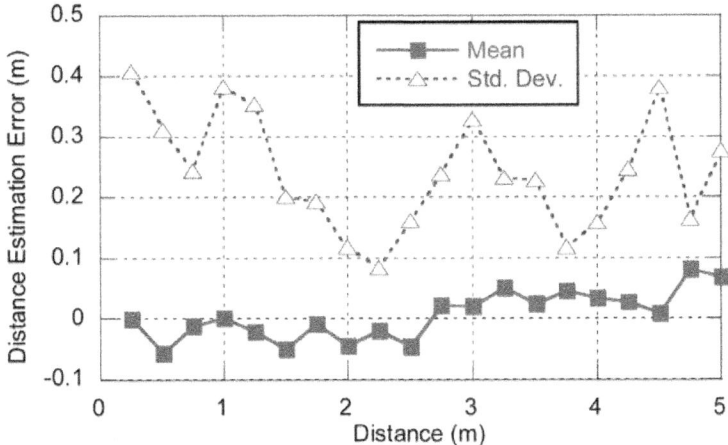

Figure 4: Distance estimation error.

Simulation Results

In this section, the system performance is evaluated and the different design alternatives are assessed. In Subsection 4.2.1 the performance of the different LT algorithms considered is evaluated using an ideal no-delay approach. The effect of delay on LT algorithms is analyzed in Subsection 4.2.2 considering generic position update latency. In Subsection 4.2.3 the specific MAC structure is taken into account, and the effect of target mobility is discussed. In Subsection 4.2.4 the different proposed strategies for acquisition and distribution of location information are assessed with the common centralized architecture with 1 LC, followed by the assessment of more advanced tracking functional architectures in subsection 4.2.5. The complete system evaluation considering all the different aspects is provided in subsection 4.2.6. Finally, experimental results with UWB prototypes are also provided.

Tracking Algorithms

The performance of each algorithm has been evaluated depending on the number of anchors used for positioning. In a first step, distance estimation and position calculation are ideally considered instantaneous, so the potential error associated with delay is not present. Concerning EKF and PF, prior simulations have been carried out in order to find the optimum values of the process and measurement noise parameters that minimized the error.

Figure 5 shows the average positioning error for a ranging residual error $\sigma_n = 0.3$ m. As it can be observed, the best performance is achieved with the particle filter. Nevertheless, the performance of the particle filter is far better than it can be expected in a real situation. The reason is that, after being optimized through simulations, the measurement error model of PF behaves almost exactly as the ranging model used to generate the distance estimation samples. Consequently, PF can deal even with highly biased measurements and the error of the particle filter decreases as the number of anchors used for location increases. In a real system, the precise characterization of the specific ranging model of the scenario would require costly measurement and calibration phases, and the use of a generic model would not provide so good results.

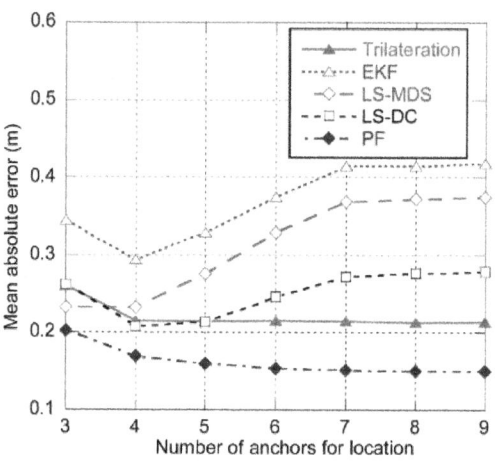

Figure 5: Positioning error. Distance between anchors = 10 m, $\sigma_n = 0.3$ m.

Trilateration, LS-MDS and LS-DC show a similar value for minimum error slightly over 20 cm, compared to 30 cm for EKF. The optimum number of anchors is four for LS-DC and EKF and three for LS-MDS. If more anchors are used, the added anchors will be more distant and will have higher ranging bias, thus increasing the positioning error. On the other hand, trilateration is almost independent on the number of anchors used to compute the position, as only the three closest anchors will be used. For every algorithm, there is an increase of the error when only three anchors are used, and the error remains constant for more than seven anchors, as the target is not likely to be in coverage of more than seven anchors.

Figure 6 shows the average positioning error for a more pessimistic value of ranging residual error ($\sigma_n = 0.6$ m). As expected, the average error is increased compared to the same configuration with ranging residual error $\sigma_n =$

0.3 m. The error increase for trilateration is especially remarkable, with results comparable to EKF in terms of minimum error. This means that trilateration requires accurate TOA estimation in order to provide good results, as it always uses three measurements for position computation and cannot take advance of diversity of measurements. Consequently, LS-MDS and LS-DC are preferred over trilateration.

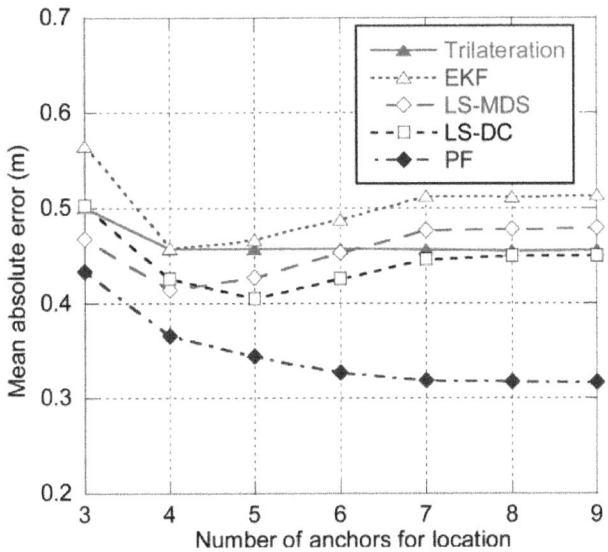

Figure 6: Positioning error. Distance between anchors = 10 m, σ_n =0.6 m.

As it was previously mentioned, parametric approaches such as EKF and PF require the accurate characterization of the target's motion model and the measurement model. With this purpose, the optimum values of the different parameters of EKF and PF that minimized the error for certain conditions (scenario layout and ranging error) were obtained through prior simulations. In order to assess the impact of the accuracy of model characterization on the positioning error, the performance of EKF and PF using parameter values that are not optimized for the current conditions is compared to the performance of the optimized filters. Figure 7 shows the average positioning error for both optimized and non-optimized EKF and PF for a ranging residual error σ_n = 0.3 m. Optimized filters use the parameter values resulting of the optimization for σ_n = 0.3 m as in Figure 5. On the other hand, in order to assess the impact of an incorrect calibration, non-optimized filters use parameter values different from the optimum. In case of the non-optimized EKF, instead of using the optimum values (resulting of the optimization for σ_n = 0.3 m), the measurement noise variance σ_e^2 is set to the value resulting of the optimization for σ_n = 0.6 m.

That is to say, we use a calibration obtained for different conditions (σ_n = 0.6 m) instead of the calibration obtained for the current conditions (σ_n = 0.3 m). As it can be observed, there is little degradation in the EKF performance, but it should be noted that the optimized EKF showed the worst performance among the different algorithms, as the Gaussian measurement noise assumption is not appropriate to model the ranging error, which is highly biased. Concerning the non-optimized PF, the value of the variance of the LOS component optimized for σ_n = 0.6 m has been used instead of the value optimized for σ_n = 0.3 m, and the mean and variance of the NLOS and NLOS2 components have been multiplied by factors 2 and 4 respectively. As expected, the non-optimized PF shows worse performance than the optimized PF, with a minimum error slightly over 20 cm, which is comparable to the performance achieved with non-parametric approaches such as LS-DC (see Figure 5). It must be noted that the parameters (mean and variance) defining the weights of each component have not been modified, although these values depend on the specific scenario and would introduce additional degradation. Finally, it should be also remarked that the difference between EKF and PF performance is mostly due to the fact that EKF uses a simple Gaussian measurement model whereas in PF we have implemented a 3-component model that is able to deal with the biased ranging error, rather than by the filters themselves. Although more complex models can be implemented by combining multiple EKF filters through multihypothesis tracking and Interacting Multiple Model (IMM) methods, the PF is better suited for the implementation of complex measurement models.

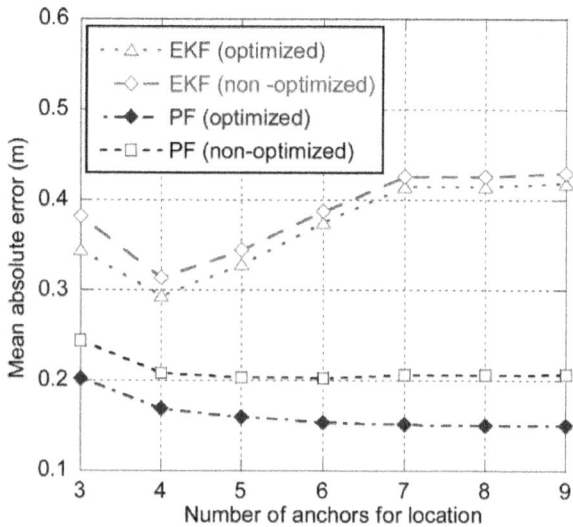

Figure 7: Impact of model optimization in EKF and PF. Distance between anchors = 10 m, σ_n = 0.3 m.

Effect of Position Update Latency

According to the position update process previously described and to the MAC superframe structure shown in Section 3.1, there are many sources of delay in the position update process: delay associated to the request and allocation of free ranging slots, duration of the ranging exchanges, transmission of the estimated distances to the location controller, latency of position computation, transmission of the updated position to the target, etc.

Delay prior to the start of the ranging exchanges has no appreciable effect on positioning accuracy. Nevertheless, delay between the ranging exchanges and the final position availability has a negative effect on positioning accuracy due to the movement of the target. We define position update latency as the time between the start of the ranging exchanges and the availability of the position at the target. Figure 8 shows the effect of position update latency on the average positioning error for the different tracking algorithms. As it can be observed, all the algorithms show a similar evolution and the error grows as position update latency increases. For 200 ms that is approximately the duration of a MAC superframe, the average error increases around 18 cm. For 400 ms (two superframes) the error increase can be as high as 45 cm. Therefore, position updates should be carried out preferably within a single superframe.

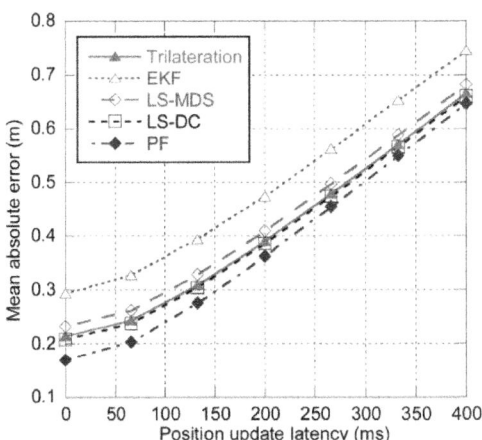

Figure 8: Positioning error depending on position update latency.

Effect of Target Mobility

Next, the effect of target mobility on the positioning error is analyzed, taking into account the timing associated to the MAC superframe structure presented in Section 3.1. Figure 9 shows the positioning error for the centralized architecture with a single LC and four anchors used for location. As it can be observed, when delays are taken into account, all the algorithms are degraded due to the movement of the target between distance estimation and position computation. Non-parametric algorithms, namely trilateration, LS-MDS and LS-DC, which are ideally independent on target speed, have a similar evolution. EKF severely degrades for speeds greater than 1.5 m/s as the estimation is based on target's previous position. Finally, although PF uses the target's dynamic model to move the particles, particles are weighted on each step according to their likelihood, so the new position is almost independent of the previous one and degradation is slightly higher than for non-parametric methods.

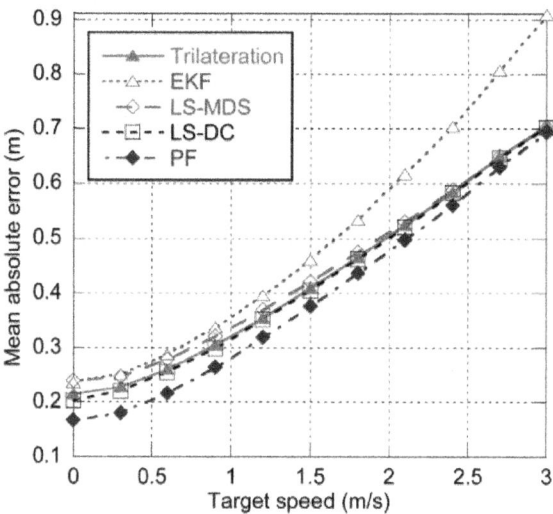

Figure 9: Positioning error depending on target speed.

Impact of Acquisition and Distribution Strategies

In order to reduce the amount of timeslots needed and consequently the latency, data acquisition and distribution enhancements presented in Section 3.2.2 can be applied. The following notation is used: SRq (Single Request), MRq (Multicast Request), SRp (Single Response), MRp (Multicast Response), NDA (No Data Aggregation), DA (Data Aggregation). The different enhancements have been simulated based on a centralized architecture with one location controller (denoted as 1LC in Figure 10) and one target. Trilateration has been used as location and tracking algorithm as its performance, without considering latency, is almost independent of the number of anchors used for location provided that more than three anchors are used (see Figure 5). This way, when the amount of timeslots needed and consequently the latency linked to data acquisition and distribution is explicitly considered, the impact of each strategy depending on the number of anchors can be appreciated better. As Three-Way ranging is used, two ranging responses are generated for each request, so three slots are needed for each ranging exchange. Concerning estimated distances transmission, in general multiple hops will be needed to relay the data frames from the target to the LC and vice versa. As it was specified in Section 3.1, there are 12 ranging slots and eight data slots per superframe.

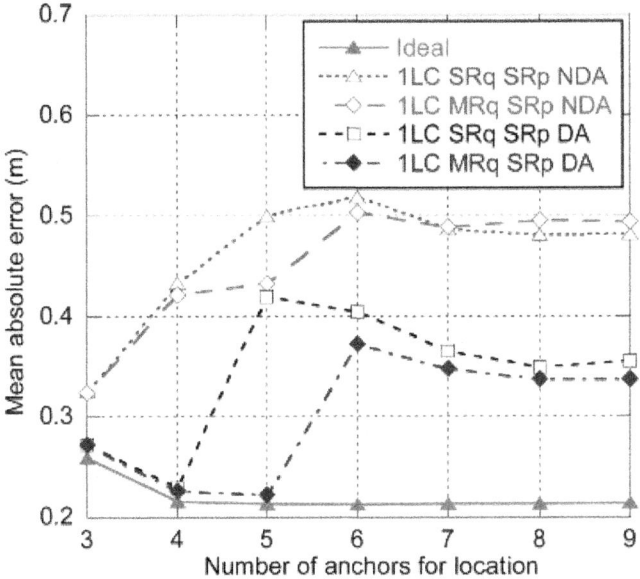

Figure 10: Positioning error for the enhanced modes.

Figure 10 shows the average positioning error for each one of the acquisition & distribution schemes depending on the number of anchors. The error for an ideal case with no delay is included as a reference. Position update latency and therefore positioning error increase are mainly determined by the number of superframes needed for ranging exchanges and estimated distances transmission. Note that results are constant for seven or more anchors as with this configuration (10 m between anchors) the target is not likely to be in coverage of more than seven anchors. The modes without data aggregation show a similar evolution as they are mainly determined by the number of superframes needed for estimated distances transmission, which depends on the number of anchors used and the number of hops from the target to the location controller. Consequently the error increases as the number of anchors increase. The only difference is for five anchors, as with SRq at least two superframes are needed for ranging exchanges, whereas with MRq ranging exchanges can be completed within one superframe and a second superframe may not be needed if the target is just one hop away from the LC.

When data aggregation is used, position update latency is mainly determined by the ranging exchanges, as location data transmission will be carried out always in a single superframe. With single request (SRq SRp DA), a second superframe will be needed for ranging when five anchors are used, so there is an important increase of the error. Then the error slightly decreases, which is related to the ratio of distances recently estimated (in the second superframe) and delayed (in the first superframe). With multicast request (MRq SRp DA) the error increase occurs when six anchors are used.

Next the effect of tracking multiple targets simultaneously on system capacity is analyzed. With that purpose, the number of anchors used for location is fixed to four and the number of targets is variable. Figure 11 shows the % of GTS slots used for location data transmission. When no enhancements are applied, the amount of slots used increases quickly and the system is eventually saturated for more than four targets, with a residual 20% left for sensor data communication. The amount of GTS slots used is reduced for the enhanced modes. As with DA a single measurement report packet is sent, capacity of these modes is limited by the availability of ranging slots to five targets (SRq SRp DA), six targets (MRq SRp DA) and eight targets (MRq MRp DA) per picocell.

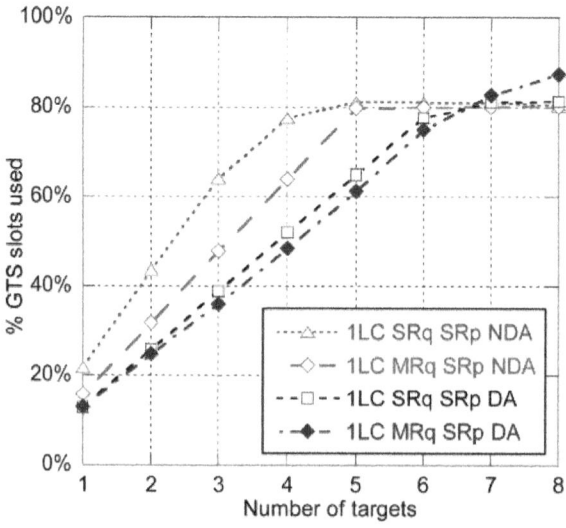

Figure 11: % of GTS slots used for location data transmission for the enhanced modes.

Impact of the Tracking Functional Architecture

In this section, the different functional architectures of the tracking system that were proposed in Section 3.2.1 are evaluated. Figure 12 shows the average error obtained for the different functional architectures depending on the number of anchors used for location. SRq SRp NDA mode has been considered. Results for the centralized architecture with one LC were already discussed in the previous section. For the distributed architecture, as the LC is implemented in an anchor neighbor to the target, location data transmission can always be done within a superframe, and latency is not determined by location data transmission, but by ranging exchanges. Specifically, as 12 ranging slots are available on each superframe and three slots are required for each ranging exchange in the SRq SRp NDA mode, up to four ranging exchanges can be done in a single superframe, and a second superframe will be needed if more than four anchors are used. Consequently there is an important increase of the error when five anchors are used, and a slight decrease for more anchors, which is again related to the ratio of distances recently estimated (in the second superframe) and delayed (in the first superframe). A similar evolution is shown for a centralized architecture with four LC as most of the time the target will be neighbor to one of the LCs. Finally, the target-centered architecture shows a similar evolution but with a slightly lower error, as there is no need of transmitting the estimated distances and the computed position, so the position update latency will always be a little bit lower.

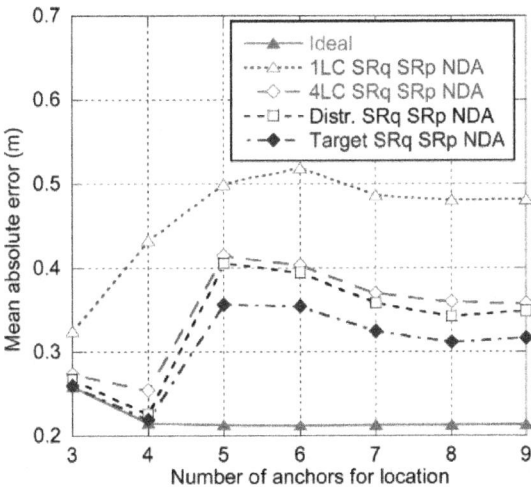

Figure 12: Positioning error for the tracking functional architectures (SRq SRp NDA).

Figure 13 shows the % of GTS slots used for location data transmission depending on the number of targets with 4 anchors used for location. As it was previously mentioned, the centralized architecture with one LC can track up to four targets until slots allocated to location data transmission are saturated. The distributed architecture and the centralized architecture with four LCs can track up to five targets until ranging slots are saturated, with a residual 10–15% of slots available for sensor data transmission. Finally the target-centered architecture can also track up to five targets but, as location data transmission is not needed, has a lower use of resources, and consequently higher capacity available for sensor data transmission. Therefore, the target-centered architecture is the optimal in terms of use of resources.

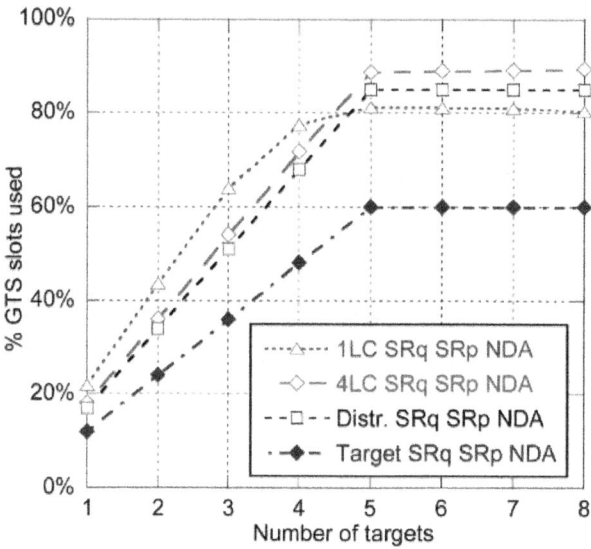

Figure 13: % of GTS slots used for location data transmission for the different architectures (SRq SRp NDA).

Complete System Evaluation

Finally, accuracy and capacity of the system are evaluated for the different algorithms considering the real MAC implementation. In order to minimize latency, the target-centered architecture centralized has been considered, together with MRq, SRp and DA enhancements. MRq has not been considered as it requires the coordination of the updates, which is complex for the target-centered architecture. Figure 14 shows the average positioning error for the different algorithms. Results for 5 or less anchors are similar than those of Figure 5 as latency is relatively small. When 6 anchors are used, there is an important increase of the error for all the algorithms, as each update will require two superframes thus increasing latency.

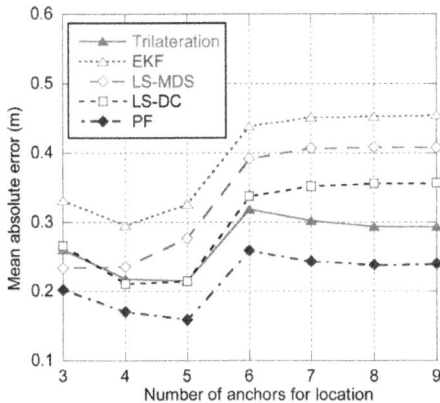

Figure 14: Positioning error for the target-centered architecture (MRq SRp DA).

In order to evaluate capacity, we have selected two options, LS-MDS with three anchors and LS-DC with four anchors, which provide an average positioning error of 23 and 21 cm respectively. Although PF provides better performance, non-parametric approaches such as LS-DC and LS-MDS are preferred over PF as they are completely independent of the scenario and do not require any prior model characterization. Again, the target-centered architecture and MRq, SRp and DA enhancements have been considered. As it can be observed in Figure 15, the system can track up to six anchors in case of LS-DC with four anchors, and up to eight anchors in case of LS-MDS with three anchors, leaving a residual 40% of GTS slots available for sensor data transmission.

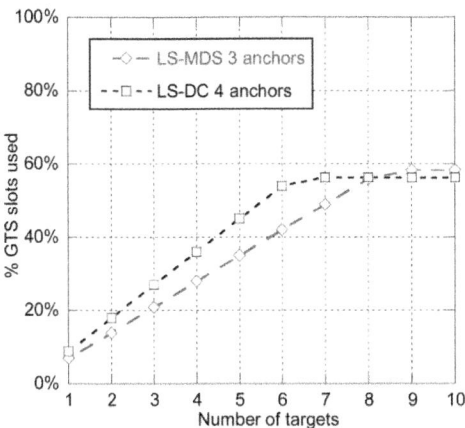

Figure 15: % of GTS slots used for location data transmission for LS-MDS (three anchors) and LS-DC (four anchors).

Experimental Results

In order to assess the accuracy level that can be reached with real UWB equipment, an experiment was carried out in a room with approximate dimensions of 5 × 3 meters using the open IR-UWB platforms already introduced in Section 3.1. These platforms are prototypes including radio-frequency, baseband and MAC hardware boards and a software MAC running on an FPGA [14]. The modulation scheme is based on Differential Binary Phase Shift Keying (DBPSK), while demodulation is performed using differential correlation between the incoming signal corresponding to the current data symbol and the previous one. The MAC layer is based on IEEE 802.15.4 and has been already explained on Section 3.1, while the main PHY/MAC parameters are shown in Table 1.

Five prototype devices have been used, one configured as picocell coordinator and anchor node, another one as target node, and the other three as anchor nodes. As the devices are started, the target node estimates the distances to each anchor node and sends them to the picocell coordinator. The picocell coordinator transmits the estimated distances through a serial RS-232 interface. Finally, a computer connected to the picocell coordinator retrieves the distances and computes the position. Consequently, the tracking functional architecture is centralized with one LC, and four anchors are used to locate a single target. LS-DC was implemented as location algorithm as it provides good performance on the different configurations simulated, is independent on target speed and is completely non-parametric, so no prior model characterization is required.Figure 16 shows a plan of the measurement scenario. The anchor nodes (green dots) were placed near the corners of the room and the estimated position was measured in 13 different locations (blue dots).

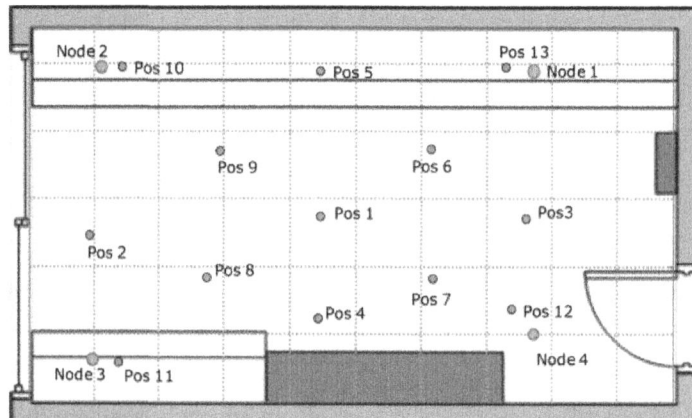

Figure 16: Measurement scenario.

Table 2 shows the mean and standard deviation of the positioning error on each one of the positions surveyed. The mean absolute error (MAE) varies from 7.8 cm when the target is in the middle of the area (position 1) to 40–45 cm when the target is in a corner of the area (positions 10, 11, 12 and 13). Standard deviation also increases for the corner placements. Average MAE is 26.6 cm, which is slightly higher than the 21 cm obtained for LS-DC in the simulations.

Table 2: Positioning error for the different positions surveyed

Positioning error	Pos1	Pos2	Pos3	Pos4	Pos5	Pos6	Pos7	Pos8	Pos9	Pos10	Pos11	Pos12	Pos13
Mean absolute error (cm)	7.8	41.6	29.4	13.6	19.4	14	15.8	17.9	20.5	44.3	44.9	37	40.2
Std deviation (cm)	2.6	4.1	2.8	6.1	2.3	5.3	7.3	3.2	10.9	11	11.8	12.5	5.1

CONCLUSIONS

In this paper, a UWB-based communication and location tracking system for Wireless Sensor Networks has been analyzed. Besides distance estimation and position calculation algorithms, the study covers some other aspects not usually considered on existing studies, such as position update latency, mobility, the functional architecture, the transmission of the associated information, and the need of tracking multiple terminals.

Concerning location and tracking algorithms, the particle filter provides the best results, although measurement model characterization would entail a costly calibration phase, and the use of a generic model would not provide so good results. LS-MDS and LS-DC provide good results and do not require model characterization. Finally, trilateration shows good results for accurate ranging measurements, but degrades as ranging error increases, and the Extended Kalman filter only works well for slow moving targets but severely degrades as target speed increases.

But, as it has been shown, positioning error is highly sensitive to position update latency. Furthermore, when combined data communication and location systems are considered, the use of temporal resources is critical, as it has an impact both on the capacity of the tracking system and the sensor data transmission capacity. In order to deal with latency and resource limitation, different solutions are proposed. The target-centered architecture is optimal in terms of latency and resources needed, as there is no need of transmitting the estimated distances to the network. For any other tracking functional architecture, the use of data aggregation is essential in order to minimize the amount of slots used for location data transmission. Finally, the number of anchors used to locate a certain target should be limited according to the number of slots allocated to ranging, so the position update can be completed within a single superframe.

ACKNOWLEDGMENTS

This work was supported by the European research project EUWB which is partly funded by the Commission of the European Union under the 7th European Framework Programme for Research and Technological Development (FP7) and here under the Information and Communication Technologies (ICT) research programme and by Gobierno de Aragón for WALQA Technology Park.

REFERENCES

1.   Liu, H; Darabi, H; Banerjee, P; Liu, J. Survey of wireless indoor positioning techniques and systems. IEEE Trans. Syst. Man Cybern. Part C 2007, 37, 1067–1080.

2.   Gu, Y; Lo, A; Niemegeers, I. A survey of indoor positioning systems for wireless personal networks. IEEE Commun. Surv. Tutor 2009, 11, 13–32.

3.   Yang, L; Giannakis, GB. Ultra-Wideband communications, an idea whose time has come. IEEE Signal Process. Mag2004, 21, 26–54.

4.   Gezici, S; Tian, Z; Giannakis, GB; Kobayashi, H; Molisch, AF; Poor, HV; Sahinoğlu, Z. Localization via ultra-wideband radios. A look at positioning aspects of future sensor networks. IEEE Signal Process. Mag 2005, 22, 70–84.

5.   Dardari, D; Conti, A; Ferner, UJ; Giorgetti, A; Win, MZ. Ranging with Ultrawide Bandwidth Signals in Multipath Environments; Institute of Electrical and Electronics Engineers: Washington, DC, USA, 2009; Volume 97, pp. 404–426.

6.   Güvenç, I; Sahinoğlu, Z; Orlik, P. TOA estimation for IR-UWB systems with different transceiver types. IEEE Trans. Microw. Theory Tech 2006, 54, 1876–1886.

7.   Sahinoglu, Z; Gezici, S; Güvenc, I. Ultra-Wideband Positioning Systems; Cambridge University Press: Cambridge, UK, 2008; pp. 63–100.

8.   Yu, K; Montillet, JP; Rabbachin, A; Cheong, P; Oppermann, I. UWB location and tracking for wireless embedded networks. Signal Process. Spec. Section Signal Process. UWB Commun 2006, 86, 2153–2171.

9.   IEEE 802.15 Working Group. Part 15.4: Wireless Medium Access Control (MAC) and Physical Layer (PHY) Specifications for Low-Rate Wireless Personal Area Networks (LR-WPANs), Available online: http://

profsite.um.ac.ir/~hyaghmae/ACN/WSNMAC1.pdf (accessed on 10 August 2011).

10. IEEE 802.15 Working Group. Wireless Medium Access Control (MAC) and Physical Layer (PHY) Specifications for Low-Rate Wireless Personal Area Networks (WPANs): Alternate PHYs, Available online: http://www. it-expo.org/docs/9_2_en.pdf(accessed on 10 August 2011).

11. Chu, Y; Ganz, A. A UWB-based 3D location system for indoor environments. Proceedings of the 2nd International Conference on Broadband Networks (BroadNets 2005), Boston, MA, USA, 3–7 October 2005; pp. 1147–1155.

12. Lo, A; Xia, L; Niemegeers, I; Bauge, T; Russell, M; Harmer, D. Europcom—An ultra-wideband (UWB)-based ad hocnetwork for emergency applications. Proceedings of the 67th IEEE Vehicular Technology Conference (VTC2008-Spring), Marina Bay, Singapore, 11–14 May 2008; pp. 6–10.

13. Wu, D; Bao, L; Li, R. UWB-based localization in wireless sensor networks. Int. J. Commun. Network Syst. Sci 2009, 2, 407–421.

14. Pezzin, M; Bucaille, I; Schulze, T; Pato, AV; de Celis, L. An open IR-UWB platform for LDR-LT applications prototyping. Proceedings of the 6th IEEE Workshop on Positioning, Navigation and Communication (WPNC'09), Hannover, Germany, March 2009; pp. 285–293.

15. Giancola, G; Blazevic, L; Bucaille, I; de Nardis, L; di Benedetto, MG; Durand, Y; Froc, G; Cuezva, BM; Pierrot, J; Pirinen, P; et al. UWB MAC and network solutions for low data rate with location and tracking applications. Proceedings of the 2005 IEEE International Conference on UWB (ICUWB 2005), Zurich, Switzerland, 5–8 September 2005; pp. 758–763.

16. Venkatesh, S; Buehrer, RM. Multiple-access design for ad hoc UWB position-location networks. Proceedings of IEEE Wireless Communications and Networking Conference (WCNC 2006), Las Vegas, NV, USA, 3–6 April 2006; pp. 1866–1873.

17. Bucaille, I; Tonnerre, A; Ouvry, L; Denis, B. MAC layer design for UWB low data rate systems: PULSERS proposal. Proceedings of the 4th IEEE Workshop on Positioning, Navigation and Communication (WPNC'07), Hannover, Germany, March 2007; pp. 277–283.

18. Macagnano, D; Destino, G; Esposito, F; Abreu, GTF. MAC performances for localization and tracking in wireless sensor networks. Proceedings of the 4th IEEE Workshop on Positioning, Navigation and Communication (WPNC'07), Hannover, Germany, March 2007; pp. 297–302.

19. Macagnano, D; Abreu, GTF. Tracking multiple targets with multidimensional scaling. Proceedings of the 9th International Symposium on Wireless Personal Multimedia Communications (WPMC 2006), San Diego, CA, USA, 17–20 September 2006; pp. 1118–1123.

20. Destino, G; Abreu, G. Improving source localization in NLOS conditions via ranging contraction. Proceedings of the 7th Workshop on Positioning, Navigation and Communication (WPNC'10), Dresden, Germany, 7–8 April 2010; pp. 56–61.

21. Daum, F. Nonlinear filters: Beyond the Kalman filter. IEEE Aerosp. Electron. Syst. Mag 2005, 20, 57–69.

22. Gustafsson, F; Gunnarsson, F; Bergman, N; Forssell, U; Jansson, J; Karlsson, R; Nordlund, PJ. Particle filters for positioning, navigation and tracking. IEEE Trans. Signal Process 2007, 50, 425–437.

23. Bai, F; Helmy, A. A survey of mobility models in wireless ad hoc networks. In Wireless Ad Hoc and Sensor Networks, 1st ed; Safwat, A, Ed.; Springer Verlag: Berlin, Germany, 2007.

24. Denis, B; Pierrot, JB; Abou-Rjeily, C. Joint distributed time synchronization and positioning in UWB ad hoc networks using TOA. IEEE Trans. Microw. Theory Tech. Spec. Issue Ultra Wideband 2006, 54, 1896–1911.

Chapter 13

CONNECTIVITY PREDICTION IN MOBILE VEHICULAR ENVIRONMENTS BACKED BY DIGITAL MAPS

Robert Nagel and Stefan Morscher

Institute of Communication Networks, Technische Universität München Germany

INTRODUCTION

As mobile ad-hoc networks gain momentum and are actively being deployed, providing users and customers with ubiquitous connectivity and novel applications, some challenges implied especially by the mobility of users have not yet been solved. Generally, it can be stated that modern applications impose higher requirements on the underlying communication solutions: more bandwidth, less packet loss, less delay and more reliability of services in terms of availability. These performance metrics are commonly termed Quality of Service (QoS). Due to the variability of node locations in mobile networks, the experienced QoS is highly time-variant. We have discussed in Nagel (2010a) that the level of attained QoS ultimately results from a proper combination of connectivity, i.e., the communication relations in a network, the chosen (and usually invariant) medium access (MAC) protocol and the traffic that is injected into the network at the nodes. If a certain level QoS is desired in a mobile wireless network, at least one of these three properties has to be actively controlled.

We have demonstrated that through controlling the amount of traffic that is injected by the nodes, effective distributed mechanisms can be employed that are, given minimal information about nodes' connectivity, able to provide (and even guarantee) a certain level of QoS. These mechanisms, however, are based on the current connectivity of the network and are effective only at present time. Should an application require a certain amount of QoS over a larger period of time, additional provisions become necessary. Although it is possible to control connectivity in certain boundaries (for instance through power control or adaptive antennas) and at a certain cost, the fundamental physical causes of connectivity themselves (location, mobility, and wireless channel state) cannot be influenced by the application as they are dictated by the user's behavior and the environment. It is, however, possible to anticipate a network's future connectivity – at least for a certain time horizon – and to compute the resulting future QoS. Upon this information, applications, services, and routing protocols could be parameterized accordingly: as an example, if the future QoS of a connection using a certain route is predicted to fall below a necessary level due to a link break, the expected remaining time until the link actually breaks could be used to proactively find and set up a backup route that uses other, potentially more stable links. Also, if a connection was to be set up for a limited time, it may be very helpful to assess if the required QoS can actually be provided by the network for the desired duration before the connection is actually established. While other work mainly uses mobility prediction in cellular scenarios to estimate hand-over times, or to support ad-hoc routing in random-mobility ad hoc scenarios, this chapterfocuses on connectivity prediction in the special case of vehicular networks.

Networked vehicular nodes can be assumed to adhere to certain rules that constitute drivers' basic behavior: they move along roads and try to avoid collisions with obstacles, such as buildings and other cars. Founded on the vehicular scenario constraint, we present an algorithm that predicts the location of vehicles based on their current state (position and velocity) and information from digital street maps obtained through the Open Street Map (OSM) project. A filter-based, self-adaptive velocity prediction algorithm is used to model the user-inflicted velocity changes. Using their current positions and the predicted velocities, possible future positions of cars on the street grid and their respective probabilities can be determined. Although the main focus of this chapter is on mobility prediction, we discuss an effective channel parameter estimation technique and propose to predict the network's future connectivity using an adaptive channel model. It should be noted that the proposed position prediction mechanism does not completely exhaust all opportunities provided by the vehicular scenario. For instance, we assume that vehicles have no information about other vehicles' missions, i.e., the planned route through the road grid. Furthermore, we make no assumptions about other vehicles' capabilities (in terms of maximum acceleration and deceleration, yaw rate, etc.). Also, we do not consider environmental properties, such as weather, street and traffic conditions, etc.

We will, however, point out and discuss the potential spots where these additional informations could be exploited to further augment the proposed algorithm. In the following Section, we present an overview of current work. Subsequently, in Section 3 we formulate the problem mathematically and in Section 4, we describe our algorithm that predicts the future positions of vehicles according to their actual state and a complemental digital road map. In Section 5, we will discuss why the channel model presented in Equation 1 is not sufficient under all conditions and present methods that can adapt to the environment. In Section 6, we discuss some simulation results. Section 7 summarizes this chapter and gives an outlook on further work.

RELATED WORK

One possible way of predicting a network's future connectivity is to use a model that reflects the individual mobility properties of a node. Given the knowledge of the initial position velocity of a node, a future position could be projected by multiplying the velocity vector with the desired time interval. Obviously, this approach does not account for changes in the length and/or direction of the velocity vector. Several more sophisticated approaches have been suggested and today, mobility prediction has become a common research topic in wireless networks.

Due to the distinct characteristics of vehicular ad-hoc networks, especially the high speeds and restricted degree of freedom in the movement of vehicles, most of the work on prediction for ad hoc networks is too general and thus inappropriate for vehicular networks. Nevertheless, some approaches are discussed here because they give an overview of mobility prediction in general. Material specific to mobility prediction in vehicular networks is very rare and the topic is often neglected in works on vehicular ad hoc networks.

Kaaniche & Kamoun (2010) presents an approach for mobility prediction using neural networks. Although it is not specifically designed for vehicular networks it should perform better than other general approaches as it is independent of the underlying mobility model. A trajectory is calculated for multiple steps in the future using several past positions in quite a similar manner as the adapting FIR filter for velocity prediction presented in this work.

However, the approach does not use any map material and hence the predicted positions may lie far off the road and may thus be unrealistic. The approach using neural networks could be used for velocity prediction in the constellation presented in this works to substitute the FIR filter, however it is expected to perform in a very similar manner and the FIR filter seems less complex to implement.

A similar approach for mobility prediction using spatial contextual maps and Dempster-Shafer's theory for decision making is formulated in Samaan &

Karmouch (2005). A framework is presented that allows prediction of the users mobility trajectory based on various bits of contextual information from e.g. user profile and map data. The approach is motivated by the fact that contextual information is becoming more common for adapting services towards the users needs and it uses the additional information in order to predict the users mobility. The concept seems feasible for e.g. cell phone users traveling on foot but does not seem appropriate for vehicular networks as the only contextual information possibly available and relevant to the future mobility is the chosen route to the destination. A complex theory to combine evidence into a prediction is not necessary in this case.

Huang et al. (2008) suggests a prediction algorithm based on fuzzy logic that aims at the prediction of a possible link break or a congested link which then triggers the construction of an alternate route. Similar to our algorithm, the prediction of a link break is based on the prediction of the future vehicle speed, the basis on which the predicted distance to the vehicle can be determined. This requires the generation of a fuzzy rule base that is then dynamically trained using Particle Swarm optimization (which in our approach is done using the adaptive filter for speed prediction). The authors use similar ideas in terms of the speed prediction but implements a fundamentally different concept. Furthermore, it is focussed on route break prediction and hence the performance of the isolated velocity prediction compared to our algorithm cannot be easily evaluated. In Boukerche et al. (2009), the authors present some general thoughts on mobility prediction in vehicular networks and propose a simple prediction algorithm based on movement vectors in order to reduce the frequency of location beacons without introducing a higher mean error in respect to the positions used for routing packets.

In Rezende et al. (2009), the same authors introduce the Network Neighbor Prediction protocol (NNP) that uses the results from their prior works to predict new routes that are going to be available in the near future and to calculate the lifetime of those routes that are currently in use. These works show in their simulation results that mobility prediction is a useful and necessary aspect in vehicular networks and should be researched in greater detail than it currently is. Another approach, although developed in the context of a different problem, is described by Althoff et al. (2010). The authors compute the set of points that could be reached by vehicles within the prediction times, given the capabilities (minimum and maximum acceleration, yaw rate, etc). of the considered vehicles. The approach is computationally complex and requires a lot of contextual information.

Using the predicted position it should also be possible to predict the future connectivity to a certain extent using an appropriate channel model.

In the context of vehicle-to-vehicle (V2V) communications, there is not yet a widely accepted channel model Paier et al. (2009). A common approach for characterizing a channel is to work out a theoretical channel model and then validate it against some appropriate measurements. Channel models are usually classified into stochastic and deterministic channel models, where deterministic channel models use ray tracing and similar techniques based on topological information about the environment in order to solve the the multi-path components (MPC) and derive a precisechannel characterization for a specific realization. Stochastic channel models, on the contrary, try to depict the statistics of the propagation channel in a more general sense that is not so much focussed on a particular situation. An intermittent approach is taken by geometry based channel models (as presented in Cheng et al. (2009)) that do use ray tracing; however, instead of using realistic modeling the calculations are based upon randomly placed objects.

In order to characterize a channel a number of parameters are used:

- The path loss exponent (PLE) α characterizes the average attenuation of the received signal.

- Large scale fading on the one hand refers to slow variations of received power due to shadowing by obstructing objects.

- Small scale fading on the other hand is caused by interference of different MPCs that result in fast fluctuations of the received power. Because these fluctuations are very hard to describe deterministically, they are usually described by means of statistics - most commonly by a Rayleigh distribution. • In order to determine how much power is carried by the respective MPCs a power delay profile (PDP) is used. The spreading of the received pulse in the time division - often referred to as the channels delay dispersion - is best described in a statistical way by the root mean squared delay spread.

- Because MPCs travel on different paths they experience different Doppler shifts. The root mean square Doppler spread describes the resulting spectrum widening of the received pulse and thus the frequency dispersion.

A large amount of research has been dedicated to the wireless channel in cellular networks. However, looking at the specifics of a vehicular channel, especially in the V2V case it soon becomes clear that its characteristics differ significantly from those of a cellular channel. On the one hand, antennas of both sender and receiver are mounted close to the ground in V2V communications, where with cellular systems usually one of them is mounted high above. This tremendously influences the propagation path of the signals and thus the channel characteristics in terms of diffraction and reflection. On the other hand, communications between vehicles commonly use the 5.9 GHz band which behaves significantly different than the 700-2100 MHz signals used in cellular systems in terms of attenuation and diffraction. Most importantly though, sender and receiver are moving at relatively high speeds in V2V scenarios, which invalidates the assumption of stationarity of the channel characteristics that is commonplace in channel models of cellular systems. That refers not only to a changing impulse response but also to a change of its statistical properties (fading distribution, PDP and Doppler spectrum) Molisch et al. (2009). According to Maurer et al. (2004), Doppler shift and Doppler spread characterize the time-variant behavior of the V2V channel mostly due to movement of the communicating vehicles and the adjacent vehicles.

This section highlights and discusses some of the works into vehicular channel modeling in the context of connectivity prediction - a topic that has not yet received much attention in literature. In Matolak et al. (2006), the authors describe a statistical V2V channel model that is restricted to small scale fading. It uses a tapped delay line model, each tap representing a multi path components received with a certain delay. Each tap has an on/off switching process modeled by a first order Markov chain allowing for persistence parameterization. In general, taps with longer delays have less probability of being on due to their lower energy. Tap amplitudes are modeled using the Weibull distribution where different parameters are proposed for different taps, based on some measurements. The authors differentiate betweendifferent scenarios, in some of which the Weibull parameters are "worse than Rayleigh" ($\beta < 2$), a phenomenon that is often called severe fading.

Maurer et al. present a geometry based IVC channel model in Maurer et al. (2004). They first try to model the dynamic road traffic and the environment adjacent to the road and then try to evaluate multi-path wave propagation through means of ray tracing. The road traffic model is based on the so called Wiedemann model and uses results from the authors previous works. As it seems very difficult to obtain real data with the necessary level of detail and the coverage, a stochastic model is utilized in order to place objects in the surroundings of the road. Different morphographic classes are defined for urban, suburban and highway scenarios that are assigned specific probabilities for different types of objects (trees, buildings, cars, bridges, traffic signs, etc.). Multi-path components are represented by rays, each of which can experience several propagation phenomena like diffraction or reflection. By calculating consecutive snapshots, a time-series of channel impulse responses can be obtained that classifies the channel for the current surrounding. The authors present measurements that validate the channel model with a standard deviation of less than 3 dB in both line-of-sight (LOS) and non-line-of-sight (NLOS) scenarios.

Paier et al. (2009) presents some measurements of V2V propagation in suburban driving conditions using GPS receivers. The authors on the one hand derive both a single slope and a dual slope path loss model from their results where the better dual slope model achieves deviations between 2.6 and 5.6 dB compared to the measured path loss. However, they find that received power is significantly less if no LOS propagation is possible. Fading on the other hand is modeled using a Nakagami distribution with variable parameters as already proposed in other works. While the distribution is Rician $\beta > 2$ as long as a LOS component is present, it turns out that fading can be "worse than Rayleigh" $\beta < 2$ once the LOS connection is lost intermittently at large distances between transmitter and receiver. Furthermore, the authors propose that the Doppler spread is dependent on the effective speed and the distance between transmitter and receiver. The dependance on distance is explained by the increasing number of scatterers at larger distances. Using this dependence, the authors present the speed-separation diagram that can help predict the expected Doppler spread and thus small scale fading characteristics at a certain distance.

In Molisch et al. (2009), the authors provide a survey on V2V channel models and measurements based on a variety of previous works on the subjects, some of which have already been discussed here. We recommend this paper as an introductory reading on the subject as it introduces important factors for channel characterization and includes a table that summarizes important parameters gathered from multiple measurement campaigns. Important aspects like environment characterization and antenna placements are also discussed that we omit here. One important result from the evaluated measurements is that at least path loss coefficients in V2V communication channels are rather similar to well-known cellular systems as long as a LOS connection is given. In terms of small-scale fading and Doppler spread, the results go alongside those presented in Paier et al. (2009).

The authors finally conclude that the amount of comparable measurements carried out on V2V channels is too insignificant in order to allow the formulation of a channel model that resembles the real-world V2V channel and important aspects such as antenna placement and shadowing by adjacent vehicles have not yet been sufficiently explored. Following this conclusion, an adequate prediction of channel quality seems challenging. Analogous to position prediction, an estimation of channel quality can be seen as a trade-off between computational complexity and prediction accuracy. An approach involvingray-tracing similar to the one presented in Matolak et al. (2006) on the one hand produces rather adequate results if provided with the necessary extent of details concerning the surrounding environment (including moving and parked vehicles), building geometries, plants and road signs. However, it seems unrealistic and infeasible to supply an on-board connectivity prediction engine with this amount of knowledge. Measurements suggest a dual-slope model for the path loss exponent as a very simple approach. Small-scale fading is usually modeled using statistical models with strong dependency upon separation distance which limits the possibilities of a prediction to a qualitative worst case approximation. Paier et al. (2009) also identifies significant differences between LOS and NLOS cases in both path loss and fading statistics.

A sophisticated approach to predict the path loss exponent using a particle filter has been proposed in Rodas & Cascon (2010), based on a log-normal fading channel model in wireless sensor networks. Particles are initialized in a random state with their respective weights being iteratively updated to provide an estimation of the path loss exponent. Weak particles with low weights are periodically replaced to avoid degeneration. The filter is parameterized with the type of the fading distribution and its variance. The authors, too, show that the PLE changes significantly as soon as the LOS is lost.

PROBLEM STATEMENT

In Nagel (2010b), we have outlined how QoS provisioning based on a network's connectivity can be attained. The basis for the computation is the connectivity matrix $\underset{\sim}{C}$ that describes the communication relations between n networked nodes. Let $\chi(x_i, x_j)$ denote the channel function, taking as parameters the physical positions x of two vehicles in the environment. A very basic channel function could then read:

$$\chi(\underline{x}_i, \underline{x}_j) = \left\{ \begin{array}{ll} 1 & \text{if } \left\| \underline{x}_i - \underline{x}_j \right\| \leq r \\ 0 & \text{otherwise} \end{array} \right.$$

(1)

This means that two vehicles i and j are connected if they are located closer than the radio range r; if they are located further apart, they are not connected. The connectivity matrix $\underset{\sim}{C}$ is then defined as:

$$\underset{\sim}{C} = (c_{ij}), \; c_{ij} = \chi(\underline{x}_i, \underline{x}_j)$$

(2)

Every node i is allowed to inject (source) traffic amounting to s_i into the network. Multiplying the source vector $\underline{s}$ with the connectivity matrix results in the load vector l:

$$\underline{1} = \underline{\underline{C}}\left(\underline{1} + \underline{s}\right) \qquad (3)$$

We have shown that the QoS criterion is fulfilled if the injected traffic is dimensioned so that each entry in the load vector $\underline{1}$ does not exceed a certain pre-defined threshold. For more detail, especially on the distributed algorithm, the reader be referred to the original paper. The problem with this approach, however, is that $\underline{s}|_{t_0}$ is only valid for the current connectivity matrix $\underline{\underline{C}}|_{t_0}$. As it is desirable to fulfill the QoS criterion over a certain time Δt, we first need to predict the future physical positions of the vehicles, estimate the channel function and then deduce the prospective future connectivity matrix:

$$\underline{\underline{C}}|_{t_0+\Delta t} = \left(\ \underbrace{\chi|_{t_0+\Delta t}}_{\substack{\text{Channel}\\\text{Estimation}}}\ \underbrace{\left(\mathbf{x}_i|_{t_0+\Delta t},\,\mathbf{x}_j|_{t_0+\Delta t}\right)}_{\substack{\text{Mobility}\\\text{Prediction}}}\ \right)$$

$$(4)$$

After that, the future source vector can be computed (Equation 3) and a decision can be made whether the current demand can be satisfied under the future network conditions and consequently, adequate measures can be taken.

MOBILITY PREDICTION

Generally, the spatial behavior of a vehicle is defined by two factors: On the one hand, speed and direction are controlled by the driver who adapts to the environment and the current situation. On the other hand, movement of a car is restricted to roads so the surrounding road topology is the major limiting factor. This is the key criterion that simplifies location prediction for vehicles compared to regular mobile users. Cars are usually not allowed to travel anywhere, they are bound to a relatively small portion of the world, the lanes. Combined with a small memory of past positions, the current velocity and direction of movement can be calculated. This further limits the amount of available future positions, as cars are usually not expected to u-turn spontaneously and velocity changes are bounded by the maximum deceleration and the maximum acceleration.

Concept

The prerequisite for the prediction is knowledge about a vehicle's current position, direction of movement and the surrounding road topology. The latter is provided by digital street maps (available, for instance, through the OpenStreetMap Project). All of these factors are very stable in terms of prediction. The destination or rather the mission of the car is assumed to be unknown to the algorithm, so at a crossroads basically all directions seem equally probable. The velocity of a car, however, is far less stable and predictable as it is directly controlled by the user and indirectly influenced by environmental factors such as traffic density, road signs and the weather. Especially abrupt speed changes are almost impossible to predict as they are often unexpected, even to the driver himself. The algorithm is sketched in Figure 1.

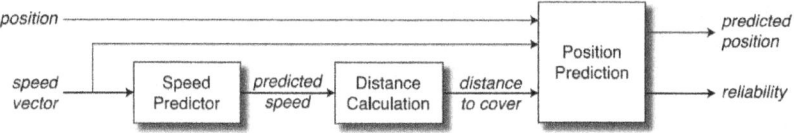

Figure 1: Algorithm Outline.

For speed prediction, we use a filter based approach that employs concepts of adaptive filters initially developed to adapt to varying channel conditions in wireless communications. Like the channel characteristics change depending on the environment, the speed change behavior of a car - or rather its driver - adapts to various environmental factors. This includes urban scenarios with steep velocity slopes and rural roads with fairly constant speeds. The character of the driver and the performance of the car also influence the prediction to a certain extent and are automatically taken into account by the adaptive filter. A self-adapting finite impulse response (FIR) filter approach based on a least-mean-squares (LMS) algorithm with relatively low depth seems ideal to adapt to both the personal behavior of a driver and the current situation. Using past and current velocities, an ideal weight vector for the past situation is calculated. Due to the low depth of the filter, the weight vector is rather unstable and consequently, it is combined with both the mean weight vector over the last iterations and a "boost" vector to improve reactivity at steep slopes. The resulting weight vector is then used to predict the future velocity, which is in turn used to calculate the distance covered in the desired interval.

The distance to cover, together with the current position and direction of movement, forms the input for the position predictor that outputs the predicted future location of the vehicle. In some cases, multiple positions are possible, for instance due to a crossroads between the current and the future position. In that case, the position that seems most probable to the algorithm is used as an output; however, internally a list of all possible locations is generated. In many situations, predominantly with cars traveling in sparsely populated areas or on highways, the prediction is rather reliable. In urban areas prediction reliability is reduced by intersections where a sudden change of direction can occur and a certain amount of past predictions may be invalidated. To make applications aware of such differences, an additional output variable was added to resemble the estimated reliability of the output.

Input Data

The algorithm requires a number of input data:

Position Data

Obviously, the algorithm requires knowledge about the actual position of a vehicle and a timestamp. The position data used in the performance analysis has been downloaded from the "GPS Tracks" section of the OpenStreetMap online portal. Selected tracks were chosen that were provided by users around the globe and thus constitute a rather broad basis of real life data. Additionally, own traces have been used. The temporal resolution of the recorded tracks was or has been resampled to one second. A statement about the spatial resolution is not generally possible as different positioning hardware from various vendors has been used for the sample data. However, we shall assume a positioning accuracy of a few meters.

Map Data

Also, the algorithm needs to be provided with map data of the area surrounding the actual position. This data, too, is provided by and downloaded from the open source OpenStreetMap project. It basically consists of an array of so-called nodes that are uniquely identified and reference a GPS position by latitude and longitude. A street is constructed by a list of subsequent nodes, forming a polyline that represents the shape of the street. Actual contiguous roads may be split apart, for instance if the name of a street changes or if two streets merge, on intersections etc.

Number of Steps to Predict

The major parameter influencing the algorithm. It is common in most parts of the algorithm and hence introduced in the high level diagram. Many parts of the algorithm also refer to it as n. Depending on the input data, the usual assumption is that one timestep equals one second. Most of the evaluations were done using a medium interval of prediction of 8 seconds - however results using different values are discussed in section 6.4.

Speed Predictor

A car typically moves in different classes of environments: urban, suburban, peripheral and highway. Each of those has different characteristics concerning the speed change of a car. On a highway speed changes are rare but usually with rather steep slopes whereas in urban areas, the speed is hardly ever constant for more than a few seconds. This allows for two different approaches in implementing a speed prediction algorithm. On the one hand, specialized algorithms could be engineered for all of the above scenarios and another algorithm that determines the algorithm that is most appropriate in an actual situation. In typical situations one would expect such an approach to give very accurate results, but clearly there are many situations where none of the implementations will be adequate.

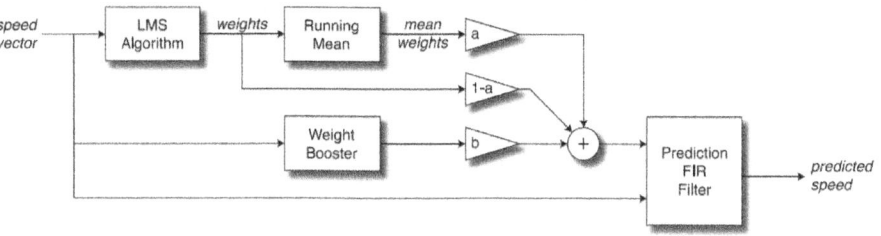

Figure 2: Speed Predictor Structure.

Furthermore, this approach involves increased efforts in development because multiple algorithms need to be designed and there is a high number of factors influencing the situation that are hard to quantize. On the other hand, it seems more appropriate to design an algorithm that automatically adapts to changing situations and as such can also adapt to factors like driver attitude and others mentioned above. This introduces some delay caused by the responsiveness of the adaption algorithm, but works also in an environment that cannot be properly classified into one of the above scenarios. In some situations, especially with quickly varying conditions, this may result in weaker performance than the approach discussed above, but the overall performance is expected to be better with less development efforts. A solution for this approach is discussed in the next sections.

Structure

The signal flow graph of the speed prediction is shown in Figure 2. The input variables are the current speed of the car and the number of steps n to predict. The only output is the predicted speed for the given time frame.

Prediction FIR Filter

The actual prediction is done in an FIR filter on the right hand side of the signal flow plan. It uses the current speed and a weight vector to predict the velocity from the last speed values. The length of the weight vector is given by the depth of the FIR filter - in the evaluations performed here a depth of 12 was used.

LMS Algorithm

The most important building block. It is the adaptive part of the algorithm and calculates an ideal weight vector from its two input values - the current velocity forms the desired signal, a delayed version forms the input for the algorithm. The weight vector is adapted with a fixed step size in the direction of steepest descent in order to achieve the minimum square error and, at the same time, limit the dynamics of the weight vector. The weight vector is recalculated each time step, for more details see Benvenuto & Cherubini (2002); Guillemin et al. (1971).

Mean Weight

Because the weight vector generated by the LMS algorithm is very reactive to acceleration and deceleration processes, it is averaged by a running mean block that calculates the mean weight vector over the length of the situation. Both the mean and the LMS weight vector are combined into a slowed weight vector by multiplication with the parameter a or $1 - a$ respectively.

Weight Booster

The LMS algorithm adapts to new situations with a delay that is roughly its depth l plus the length of the prediction interval n, which equals the number of memory elements involved in the adaption. A change in velocity needs to pass through most of the memory elements before its effect becomes visible in the weight vector. For instance, for a car traveling in a city, the weight vector produces rather stable results while the car is traveling at a constant speed but it will react slowly to sharp braking or fast acceleration.

Table 1: Speed Prediction Parameters with example values used during development

parameter	example value	description
n	8	number of steps to predict
l	12	depth of FIR filters and dimension of weights vector
a	0.275	influence of the mean weights vector
b	2	influence of weight booster
c	0.2	boost limit
d	1	boost gain

The algorithm is designed to fix this problem by manipulating the weight vector in order to emphasize the most recent speed history elements to react more quickly to a spontaneous change in behavior: a length l base vector is multiplied with a scalar calculated from the slope of the velocity curve and is bounded above by the boost limit c. In order for the impact of the booster to remain present for a longer period, the generated "impulse" is broadened using a unity-weight FIR filter.

Parameters

The performance and precision of the speed predictor depends on some fundamental parameters that are summarized in Table 1. The given values are the result of some evaluations during the design phase based on few exemplary scenarios and should give a rough idea to start an implementation. However for a proper implementation a more thorough, numerical optimization is recommended but out of scope of this essay. It is important to note that all of the below parameters influence the prediction in a way that usually makes adoptions to all parameters necessary if one parameter is changed. In many cases more than one possibility exists that can lead to a desired result for one scenario, but looking at multiple scenarios usually only one if any of the possibilities lead to an overall improvement of performance.

Number of Steps to Predict (n)

This key parameter determines the number of steps to be predicted — $n = 8$ means the algorithm predicts the speed in 8 time steps. Obviously a higher value increases the prediction error, whereas lower values gives more precise predictions. The setting of this parameter is very important because its influence on the other parameters is tremendous, for instance a high value for n will on the one hand require a higher a and on the other hand require more influence of the weight booster, b. Also, this parameter is common with all components of the algorithm, so its influence has to be regarded globally. Different settings and their impact, especially on position prediction, are discussed in section 6.4.

Depth of FIR Filters (l)

The FIR filters' depth used in the speed predictor is a common value because all blocks share the weight vector. Also the parameter l is, unlike all other parameters mentioned here, a design time parameter that cannot be changed easily as it is hard-coded into the FIR filters and the constants. Nevertheless, its influence on the prediction should be discussed here. For the fact that the depth of a filter resembles its amount of memory elements, higher values for l give more stable and less reactive prediction results. Changes in the situation need more time to propagate through the memory elements, hence it takes a longer time to adapt to changes. Smaller values for l improve reaction time but also result in less stable and more fluctuating predictions that often overshoot at slight changes.

Influence of the Mean Weights Vector (a)

The weight vector in the standard case (disregarding the weight booster) is combined from the current weight vector produced by the LMS algorithm and its running average. Setting a to the maximal appropriate value a = 1 produces a very stable weight vector but also removes the direct influence of the LMS algorithm to the weight vector and thus the reactivity. This is caused by the fact that in this model, the mean weight vector is never reset and thus provides an "all time average".

Influence of the Weight Booster (b)

This parameter determines the overall influence of the weight booster. Higher values tend to produce overshoots as a trade-off to slow response to a change in situation if lower values are used. Generally all three values influencing the weight booster should be tuned according to the length of the prediction n. With high n, b should be increased because a faster reaction is necessary due to the latency of the LMS algorithm.

Boost Limit (c)

The "boost" vector, or more precisely its scalar values, are influenced by the slope of the velocity curve. The "boost limit" defines an upper bound to those scalar values.

Boost Gain (d)

This parameter multiplies the influence of the slope difference before the broadening and limiting of the pulse. Thus a higher value generates very quick increase once a steep slope is detected - in other words, it pushes the boost weight vector more quickly to the limit. Lower values produce a smoother response to steep velocity slopes.

Distance Calculation

The speed predictor predicts the vehicle speed some time steps ahead. The position predictor in turn requires as an input the distance to cover in the next time steps to calculate the future position. The most precise approach is to predict a velocity value for each time step in the prediction period and sum up the difference. Because this requires a set of n speed predictors which increases the computational efforts by n, a simpler approach is chosen in this implementation. Our algorithm uses the current speed and the predicted speed and calculates a linear approximation between the two. The distance to cover s is then the area under the speed curve for a duration of t (n time steps):

$$s = t \cdot \left[v_{current} + 0.5 \cdot \left(v_{pred} - v_{current} \right) \right]$$

Position Predictor

The position predictor uses the current position and direction of movement, digital map data as well as the predicted distance to cover as inputs and outputs a predicted position and its reliability. It is invoked once per time step and tries to first find the current road segment of the vehicle, then determines a number of possible prediction paths and finally chooses the most probable path and returns its end point.

Determine Current Road Segment

All known nearby road segments (taken from the digital map) are evaluated for the distance of the current position to the closest point on the respective road segment. Three criteria must be met in order for a road segment to be chosen:

- The distance to the closest point is smaller than a threshold.
- The absolute value of the difference between the direction of movement and the road segment's direction is not larger than $\pi/2$ because vehicles usually do not move perpendicular to streets.
- It is the closest road segment satisfying both criteria 1 and 2.

In case no road segment is found that fulfills all of the above criteria, the algorithm returns the current position as prediction result with an estimated accuracy of 0%. Possible causes range from wrong GPS positions during the initialization phase of the GPS device and inexact map material to driving or parking on streets or private property that is not (yet) included in the map material.

Determine Possible Paths

First, the remaining distance from the current position to the respective end point of the road segment sr is calculated. If the road segment's end point is further away than the distance to cover (sr ≥ s), the predicted position will be located between the current position and the road segment's end point. Therefore, the predicted position is determined along the road segment's polyline towards the end point, covering the given distance s. In the case that the remaining distance sr is smaller than the distance to cover (sr < s), the predicted position is moved to the road segment's end point and that distance is subtracted from the remaining distance. Subsequently, the next road segment of the prediction path is determined. If the mission of the vehicle is known in advance, the next road segment is chosen according to that mission. Otherwise, in order to find the next road segment, the number of possibilities is determined from the digital map: at a junction, all connected street segments are considered possible candidates. The current road segment, however, is not considered as an alternative — in other words, the vehicle is not expected to u-turn. Three cases exist:

- No candidates exist, so the current road segment ends in a node that has no other road segments referenced. In this case the relative probability of the current path is decreased in the relation to the amount of distance already covered.

- One candidate means there is no choice and the vehicle is moving along a road without an intersection at the current node. Determining the next road segment and updating the path accordingly is trivial.

- Multiple candidates are available, so the predicted path hits some kind of intersection. Hence the process of determining the next road segment becomes a bit more complex: Initially, all candidates must be assumed to be equally probable.

The procedure is repeated recursively until all distance s is covered and all possible paths of length s have been determined (effectively yielding a tree of possible road segments, with leaves at all possible future locations).

Pick Best Path

It may be desirable for an application that the prediction comprises all possible future realizations. However, if the prediction routine should return the future position along only one predicted path, the best of the alternative paths found must be chosen. If the mission of an observed vehicle is unknown to the algorithm, it must more or less issue a guess as to what option the driver of a car will go for. The range of alternatives is narrowed down in three steps:

- Estimated probability: In the current implementation of the algorithm, this first step will only remove the paths that end in a dead end and hence have a reduced relative probability. All other paths are considered equally probable and hence cannot be classified by their probability. For instance, the car hits an intersection with three alternatives, one of which being a dead end street. The dead end would be removed from the candidates, whilst the other two possibilities are equally probable.

- Way Changes: The number of street changes is the primary decision criterion for the algorithm. It is assumed that in case multiple paths exist, the driver stays on the current street. Hence the path with the least number of street changes is favored for the prediction. It is furthermore assumed that if it is necessary to change the road at some stage in all paths, the driver still stays on the current road as long as possible.

- Direction Difference: Should the way change criteria be unable to choose one candidate, the total difference of direction along the path is considered. Assuming the driver to be lazy, the path encountering the least change in direction is chosen to be the best path.

Clearly, criteria (b) and (c) do not increase the probability of a certain path. These are merely decision criteria in order to choose one path from multiple options. Choosing a random path statistically produces the same error, but has a severe disadvantage in terms of continuity: as the algorithm is executed each time step, it should return consistent values from one step to the next; when using a random selection, it is most likely that the algorithm will return a completely different position each time it is invoked. From a statistical point of view, this does not change much but for another program or algorithm that is based on the results of the prediction it may very well change things depending on the application. For the very same reason, it is very important for the algorithm to return the estimated probability of a given prediction because another program can then classify the prediction accordingly.

CHANNEL PREDICTION

It is clear from Equation 4 that the network's future connectivity does not only depend on the future vehicles' positions. An adequate channel model has to be chosen and constantly updated to reflect the changing radio environment. In Equation 1, we have shown an exemplary simple channel function, the disc model. We shall call two nodes i and j connected if the path loss $\beta(d)$, a function of the distance between the two nodes, does not exceed a certain threshold β_t :

$$\beta_t \overset{!}{>} \beta(d) = 20 \log_{10} \frac{4\pi d_0}{\lambda} + 10\alpha \log_{10} \frac{d}{d_0}$$

$$(5)$$

The path loss consists of two components: a constant added that reflects the loss related to the wavelength of the signal and a distance-dependent term that represents the propagation of the radio wave through space and the resulting diminishment of power due to the growth of the wave sphere's surface. Due to various propagation effects, the path loss exponent (PLE) α is variable, to reflect various environments' radio properties (and usually ranges between 2 and 3). To account for reflections, scattering and shadowing, an additional stochastic variable β_s is introduced that is log-normally distributed with zero mean and variance σ^2:

$$\underbrace{\beta(d)}_{\text{measured}} = C + \underbrace{10\alpha}_{\text{unknown}} \log_{10} \underbrace{\frac{d}{d_0}}_{\text{known}} + \underbrace{\beta_s}_{\text{unknown}} \tag{6}$$

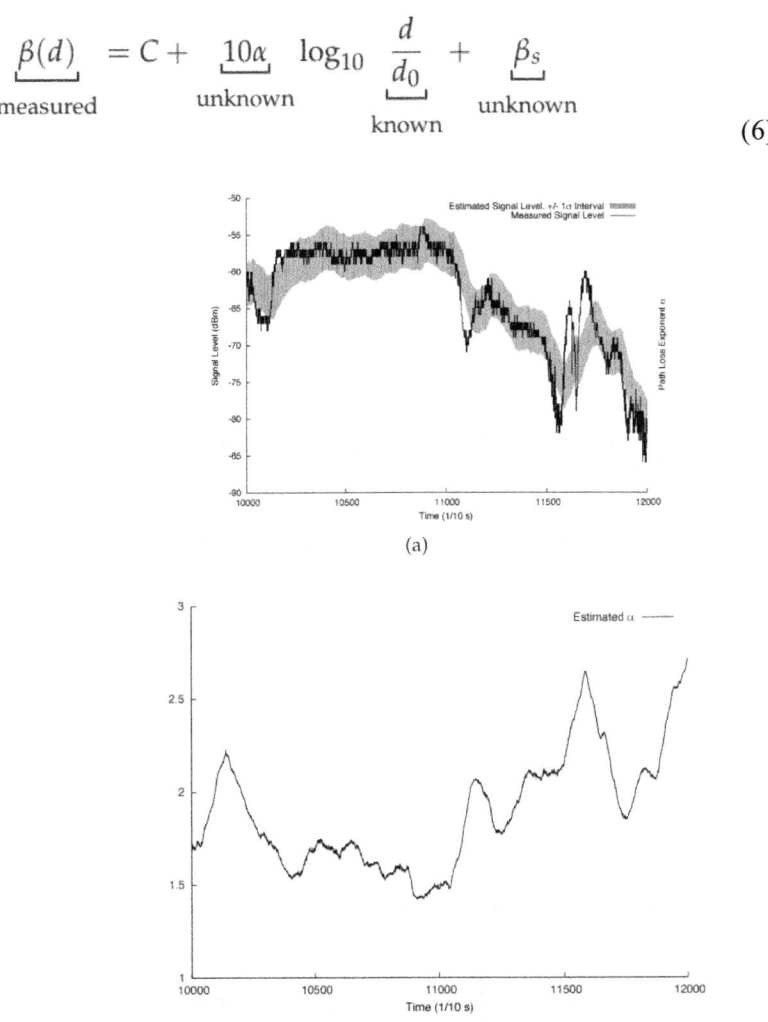

(a)

(b)

Figure 3: Parameter estimation using particle filter: measured and estimated signal level, estimated PLE.

Through constant exchange of position messages, two nodes can determine their distance d and at the same time measure the path loss β(d) between them (terms marked as "known"). Assuming that βs is log-normally distributed and knowing the order of magnitude of the variance, we suggest to use a particle filter for online estimation of the PLE α and subsequent prediction, analogous to the work presented in Rodas & Cascon (2010). To study the vehicular channel, we have recorded and evaluated several hours of measurements. Figure 3(a) shows the measured path loss (solid line) and the path loss computed using the estimated PLE plus βs's 68% (one standard deviation) confidence interval (filled curve). The standard deviation of the measured signal level was estimated around 3 dB, the system constant C was -42 dB and the duration of the displayed dataset is 20 seconds. The estimated PLE is shown in Figure 3(b). For connectivity prediction, we propose to employ the same concepts as used for position prediction to at least estimate the trend of the PLE. Furthermore, as we have discussed in the section on related work, it is very important to distinguish between LOS and NLOS conditions. In Nagel & Eichler (2008), we have introduced a method for V2V channel simulation in environments that include objects that possibly obstruct a direct LOS path and have discussed how a dual-slope channel model can be implemented to account for these objects. Consequently, we propose to complement the path loss formula in Equation 6 accordingly and incorporate the information on buildings and other obstacles from the digital map material (that is already used in the position prediction) to determine if the path between the predicted future vehicle positions is LOS or not.

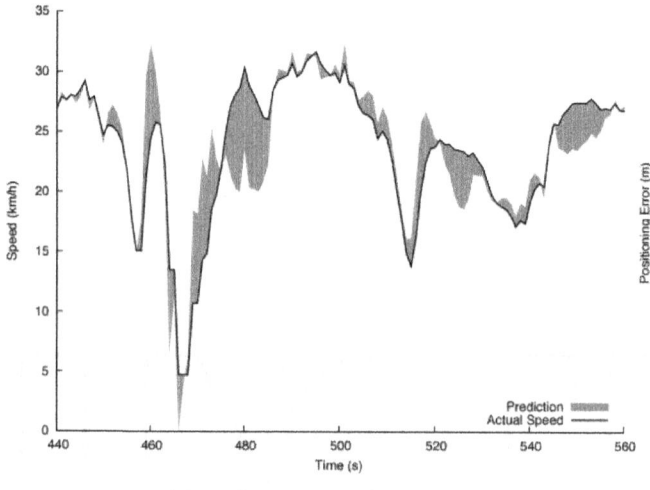

(a) Velocity prediction

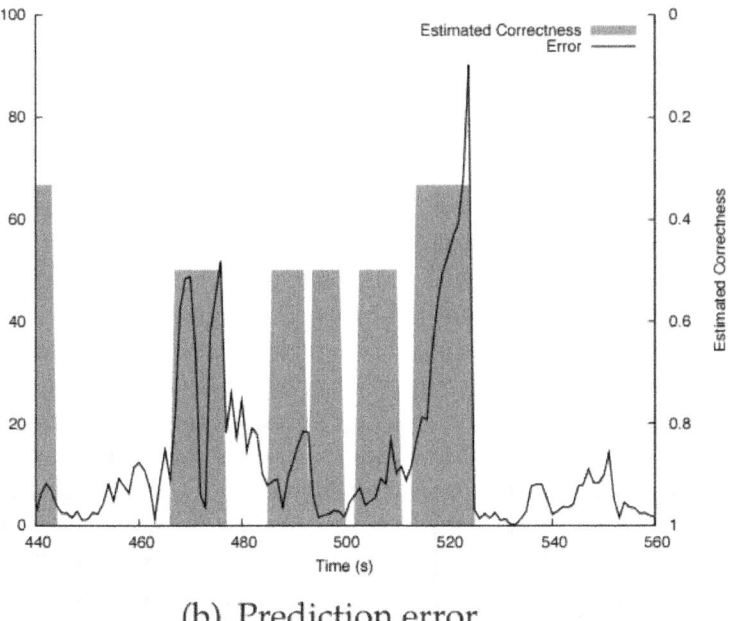

(b) Prediction error

Figure 4: City Scenario.

The propagation breakpoint derived from the map should then be accounted for in the PLE estimation. In simulations, we have realized quite accurate channel parameter estimations using a particle filter that accounts for LOS and NLOS conditions, based on information about surrounding radio obstacles. It is, however, clear that the effects of large-scale fading have a large impact on the future connectivity but are hard to account for. Due to positioning errors, it is virtually impossible to predict small-scale fading effects. When evaluating the connectivity matrix computed from the predicted positions and the predicted channel, additional information about the reliability of the prediction should be provided and accounted for.

RESULTS

Three representative scenarios were chosen for the performance evaluation of the developed algorithm. These scenarios are based on GPS tracks downloaded from the OSM portal that were selected to provide maximum diversity in the results presented below: the chosen tracks were recorded in a city as well as in suburban and highway surroundings. We have evaluated the three scenarios concerning the accuracy of both the speed prediction and the resulting predicted position under the three different environmental settings.

City Scenario

The first data set represents a typical city scenario. It has been recorded in the german city of Herne in the Ruhr area, with speeds of up to 50 kilometers per hour and a total length of about 20 minutes.

Figure 4(a) shows a section of the actual vehicle speed (solid line), the area of the filled curve reflects the velocity prediction error where the upper or lower edge of the area marks the predicted speed. For easy comparison, the predicted values are shifted by 8 seconds, so that the real value and the value that has been predicted for that instant are matched in time. The results show that especially at steep slopes, the algorithm overshoots significantly and predicts too low or too high velocity values. This could be tuned using the parameters of the speed prediction in order to achieve better performance in the particular scenario, but the impact on other scenarios is hard to estimate and thus requires significant research efforts. Figure 8(a) shows the distribution function of the position prediction error in meters (upper blue line). The mean error is 13.4 meters, the median amounts to 7.5 meters. Figure 4(b) shows a section of the prediction error over time; the filled curve represents the estimated probability (correctness) of the prediction. Note that the second Y axis has been reversed for better readability.

As explained in section 4.5, the estimated probability generally equals 1 if the position predictor identifies only one possible path for the vehicle, based on the map data. In the presented case that the mission or route of the car are unknown to the algorithm, the choice of the path used for prediction is arbitrary once it encounters p multiple possible paths. Hence the estimated probability drops to 1/p. This, in turn, means that if a large error occurs while the estimated probability is 1, an error in the position predictor or the map material should be assumed, while high prediction errors with low estimated probability are most likely to be produced by the fact that the mission of the car is unclear. Should the position predictor be either unable to find the current segment or to find any path to continue, the estimated prediction probability drops to zero. This is the case at the beginning of the city scenario and is the result of pulling out of a private parking area in a sort of backyard. As there is no map material covering that area, the position predictor cannot give useful predictions based on the fact that it is unclear on which road the vehicle is driving. In such situations, the position predictor simply returns the current position.

To get an idea of the nature of an error, it is helpful to visualize the real and the predicted path as shown in Figure 5. The actual path is shown as a black solid line while the predicted path is shown as a dashed orange line.

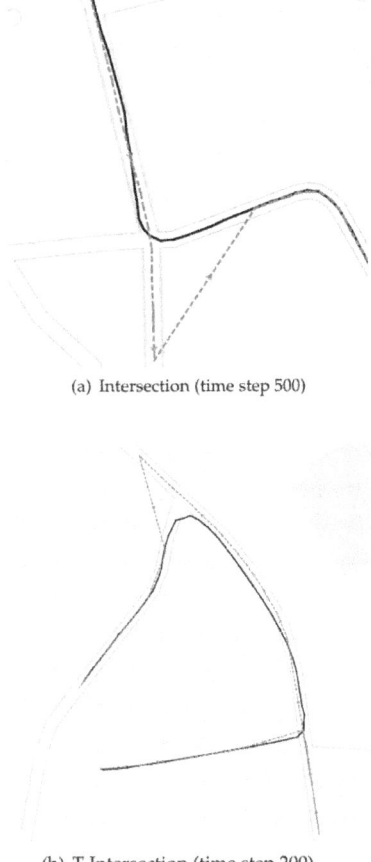

(a) Intersection (time step 500)

(b) T-Intersection (time step 200)

Figure 5: Scenario 1 - Prediction Behaviour at Crossroads.

Around time step 520, Figure 5(a) shows a typical situation in which the prediction algorithm encounters a crossroads. For the reasons explained in Section 4.5, the algorithm always prefers to choose the path that continues on the current road. This can be seen in the Figure where the predicted green dots continue straight on, while the real, blue trace turns onto the intersecting road. Once the car is on the new road, the prediction adapts to the new situation and continues its prediction along the new road. This can also be seen in Figure 4(b), where the error peaks due to the large discrepancy between the real and the predicted position. At the same time, the estimated probability drops to 0.33, due to the fact that the algorithm recognizes three alternative paths at the intersection. Figure 5(b) shows a similar example around time step 200 as the car moves towards a T-crossing. The algorithm chooses the path with the lowest total change in angle, which in this case is the wrong choice. This, again, results in a drop of the estimated probability and a peaking error. As can be seen from the examples above, the high error in urban scenarios is widely based on the fact that usually many intersections lie along the path and high prediction errors are introduced if the algorithm chooses the wrong path. We have argued that this fact can be significantly improved if the cars mission is known to the position predictor and, consequently, the correct path can be used for the prediction.

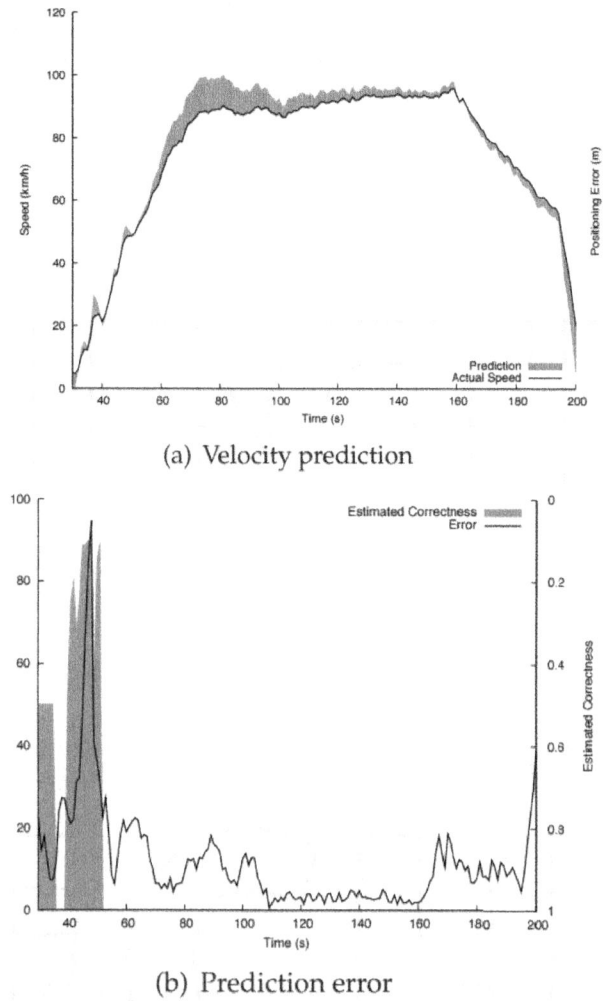

(a) Velocity prediction

(b) Prediction error

Figure 6: Suburban Scenario.

This also implies that the mean prediction error is not a very good metric in order to assess the precision of the prediction. The algorithm may predict the cars position with an error of less that five meters in cases where there are no intersections, but at each intersection an error of up to 50 meters is possible that will greatly influence the mean error. A more suitable metric to identify the amount of such situations in the scenario is the discrepancy between the mean error and the weighted mean error that includes the estimated probability. The weighted mean error is simply the mean over the prediction error, multiplied with the estimated probability. Because the estimated probability drops at intersections, a prediction error in this situation is weighted less than an error occurring along a straight, intersection-free road. The weighted mean error in this situation amounts to about 7.4 meters, which is significantly less than the mean error of 13.4 meters and therefore shows that the prediction quality could be greatly improved with knowledge of the cars future route.

Suburban Scenario

The second scenario covers the suburban area of Wiener Neustadt in Austria. The driver first hits the B17 road and afterwards enters the suburban area where the car is parked at a shopping center. The section of the velocity graph in Figure 6(a) shows a short drive to the highway like state road. The speed prediction is rather stable whilst traveling at constant speeds between second 50 and second 200. This is accompanied by a rather small prediction error as shown in Figure 6(b). A peak in the prediction error at second 40 occurs due to a wrong choice of the next segment - again based on the fact that the cars mission is unknown. The fact that the algorithm had the choice of multiple directions is visualized by a significant drop of the estimated correctness curve. Due to a very sharp deceleration around second 200, Figure 6(b) shows another peak without a drop in probability for the fact that no other possible directions are identified. The reason for the braking is unclear, an explanation cannot be found in the map material nor the satellite image of the area. Figure 6(b) also shows a few more peaks based on decisions for the wrong directions and braking actions. The distribution function shown in Figure 8(b) (upper blue line) is slightly flatter than the one from scenario 1. The median error (about 10 meters) is slightly higher than in the city, probably caused by the fact that either map material or the GPS device used to record the track are less precise than in the city scenario. The 90% percentile is significantly smaller (22 meters compared to 34 meters), which supports the before assumption and leads to the conclusion that the overall position prediction is better in the suburban scenario. The mean error of 13.5 meters, however, is almost identical to the city scenario.

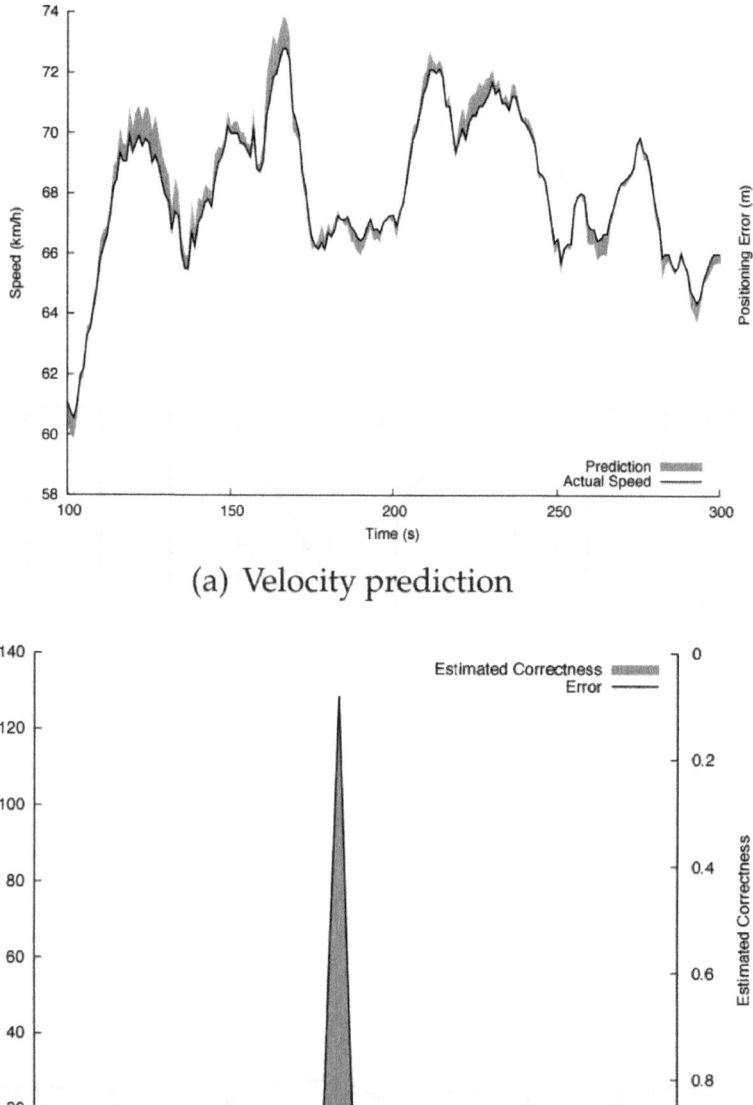

(a) Velocity prediction

(b) Prediction error

Figure 7: Highway Scenario.

Highway Scenario

The third scenario covers a ride on a highway near the English town of Cambridge during rush hour traffic, which explains the unsteady velocity curve shown in Figure 7(a). Again, a representative section has been chosen for presentation. The prediction error plotted in Figure 7(b) shows constantly low prediction errors. From the distribution function shown in Figure 8(c) (upper blue line) reveals a very steep slope, which means excellent performance of the algorithm as more than 95% of all errors are below 10 meters. The mean error amounts to only 4.6 meters and the median is 3.8 meters. One notable peak of the error, accompanied by a drop in the estimated probability occurs around time step 800 (see Figure 7(b)). At the instant when the car is crossing a highway bridge the algorithm wrongly chooses the current road segment to be part of the road below the bridge that shares a node with the highway. Consequently, the algorithm chooses the wrong road segment to continue prediction. This error is caused by an unfortunate combination of a small measuring error of the GPS device and an imprecision of the map material, where the highway shares a node with the intersecting road (although they are on different levels).

Another error source stems from the fact that in the used map material, a road is represented by a polyline, and all vehicles are assumed to be positioned on this line. In reality, highways consists of a number of parallel lanes that have a certain lateral displacement. Although a suitable representation of road lanes has been suggested, the data available today does not yet represent different lanes. We expect, however, that the error would be decreased if this information could be accounted for in the prediction.

Influence of Prediction Length

The results that we have discussed above were obtained through prediction with period lengths of eight seconds. This sections evaluates the same scenarios with longer prediction lengths of 16 and 32 seconds. The resulting position error distribution functions are shown in Figure 8. The steepness of the distribution functions decrease as the prediction length increases, due to less prediction accuracy across a longer period of time, i.e., the tendency to produce greater errors. The maximum error also increases dramatically, because in case that a wrong path is chosen, the prediction continues along the wrong path for a much longer time before it is corrected as the car turns the other way. The mean error is also shifted towards higher values with increasing prediction intervals.

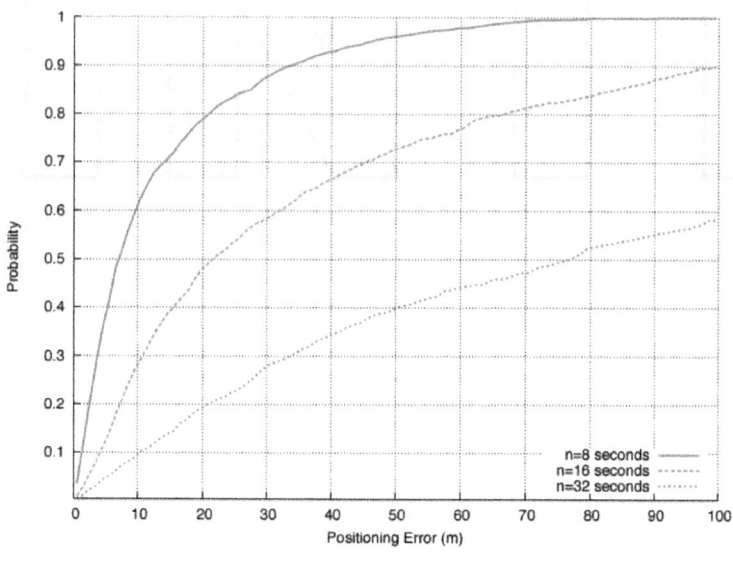

(a) Scenario 1 - City

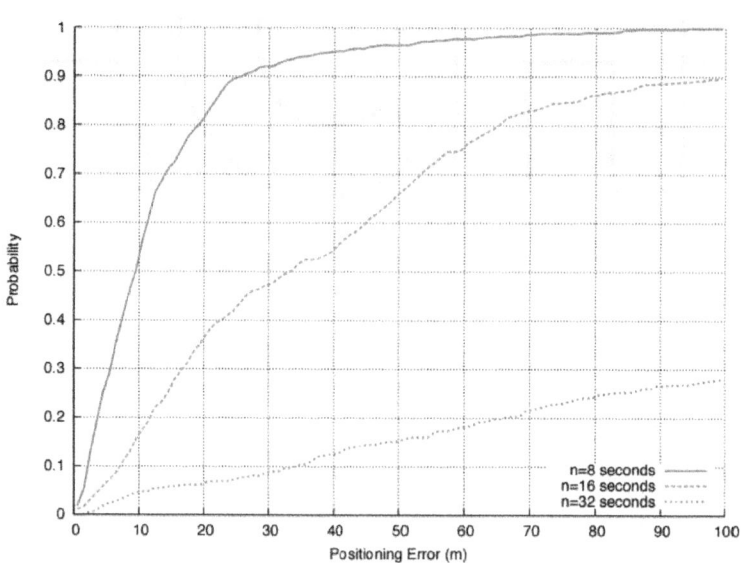

(b) Scenario 2 - Suburban

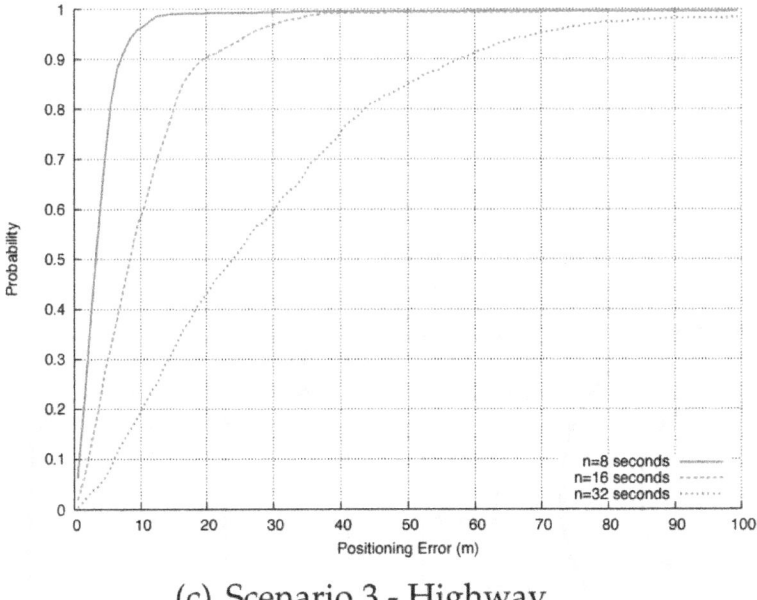

(c) Scenario 3 - Highway

Figure 8: Prediction Error Histogram - Different Prediction Lengths.

In the city and suburban scenarios, the prediction is already significantly less reliable with n = 16 time steps (i.e., seconds) as a prediction interval. It is rendered basically useless with a prediction interval of n = 32 and the majority of errors are out of scale of the histogram. The highway scenario, however, behaves much more stable as the prediction interval is increased and still returns useful results using a prediction horizon of 32 seconds. To a large extent, this is based on the relatively stable velocities and absent alternative paths along the way.

Path Loss Estimation Error

To determine the future connectivity of a network, a node has to predict the positions of all relevant vehicles (usually the one- or two-hop radio neighborhood) and determine the resulting path loss to decide whether it will be connected to this vehicle or not. The first error source of the path loss estimation is, of course, the error induced by inaccurate position prediction. Fortunately, this error term strongly depends on the distance d between the two involved vehicles. Let us assume that the estimating vehicle has perfect ego positioning and position prediction and let Δd denote the maximum positioning error. The resulting absolute estimation error range is then:

$$\Delta\beta(d, \Delta d) = \beta(d + \Delta d) - \beta(d - \Delta d)$$
$$= 10\alpha \log_{10}\left(\frac{d + \Delta d}{d - \Delta d}\right)$$

(7)

The second influence on path loss estimation is the accuracy of the predicted path loss exponent. Let $\Delta\alpha$ denote the PLE's maximum estimation error. The resulting absolute path loss estimation error is:

$$\Delta\beta(d, \Delta\alpha) = \beta(d, \alpha + \Delta\alpha) - \beta(d, \alpha - \Delta\alpha)$$
$$= 20\Delta\alpha \log_{10}\left(\frac{d}{d_0}\right)$$

(8)

Clearly, there exists a strong negative correlation between the path loss estimation error and the distance to the tracked vehicle, d. With increasing d, the implications of prediction errors become less important with respect to the path loss. The situation is contrary regarding the PLE estimation: because the relation is linear, a large path loss estimation error results if the distance d to the tracked vehicle increases. The problem is that Δd and $\Delta\alpha$ are not known at runtime; therefore, we suggest to keep the prediction results and constantly compare them against the predictions to obtain a statistic of the errors. This information should consequently be used to determine a prediction's reliability.

CONCLUSIONS AND OUTLOOK

In this chapter, we have presented an algorithm for the self-adaptive prediction of mobile nodes' future positions. The algorithm is targeted at vehicular applications with nodes that move along the road grid, of which a digital map is available at runtime. We have introduced the necessary building blocks along with their parameterization, discussed some performance studies and pointed out individual strengths and shortcomings. Three exemplary scenarios have been studied: city, suburban, and highway. Given a prediction interval of eight seconds, the algorithm performed well in all of the scenarios, resulting in a mean prediction error of only about 14 meters. On a highway, the mean error is less than 5 meters. As the prediction interval increases, the performance of the algorithm degrades significantly in the city and suburban scenario. On a highway, however, the mean error is around 30 meters for an interval of 32 seconds which may still be acceptable, depending on the application. We have already argued in the discussion that the position prediction accuracy in the city and suburban scenario is mainly degraded due to an incorrect path selection as the considered vehicle approaches an intersection. In our studies, the future path has been selected randomly from the set of possible paths.

The necessary assumptions have been explained in Section 4.5.2. If the information is available, we strongly propose to consider vehicles' missions for path prediction. Considering the city scenario, if only those prediction errors are evaluated for which the path selection is correct (i.e., the predictor estimates the correctness as 1), the mean error can be decreased to about 8 meters, the median to about 5 meters. Taking the mission into account, these extremely low errors seem feasible. The computation of the distance to cover is currently calculated using the area under a linear graph between the current velocity and the predicted velocity, thus assuming a linear acceleration. Some thoughts should be given to a substitution of this simple approximation with a more sophisticated implementation.

One idea that is rather complex in terms of computational efforts is to use n velocity predictors to predict a velocity for each time instant and thus removing the interpolation. Another approach could be to model the acceleration and deceleration behavior of a typical driver. If the vehicle is autonomous, reproducing the design of the longitudinal controller could increase the prediction performance. Velocity prediction, too, offers some optimizations opportunities. The key measure necessary here is a numerical optimization of the parameters a, b and c mentioned in Table 1 over a large number of scenarios of adequate length. Appropriate parameter sets could be computed beforehand (and even optimized online) and information about the current driving situation could be used to select the most suitable set.

This selection, in turn, could be used to provide other applications with valuable information about the current environment. It is expected that an adoption of these parameters will lead to a somewhat significant improvement of the speed prediction. An urban scenario requires much quicker reactions to speed changes and thus needs more contributions and stronger influence of the weight booster than a highway scenario. The highway scenario, in turn, profits from a more stable prediction based to a large extent on the mean weight vector and requires virtually no influence of the weight booster. Longer-term prediction of the wireless channel still imposes the largest problem when it comes to evaluate the future connectivity from predicted positions. Further work is necessary to evaluate the dynamics of the radio channel and design an appropriate predictor. We propose to include further information about the environment (see above) in order to distinguish between different surroundings and consequently adjust the appropriate channel parameters (such as the variance of the large-scale fading). An interesting idea in this context is to share and aggregate knowledge of the communication channel obtained from measurements between nearby vehicles. Another situation that requires attention is the channel prediction for vehicles that are actually outside of a node's communication range and channel parameter estimation is obviously not possible. In this case, the channel has to be estimated from measurements conducted with and from nearby vehicles.

REFERENCES

1.   Althoff, M., Stursberg, O. & Buss, M. (2010). Computing reachable sets of hybrid systems using a combination of zonotopes and polytopes, Nonlinear Analysis: Hybrid Systems 4(2): 233 – 249.

2.   Benvenuto, N. & Cherubini, G. (2002). Algorithms for Communications Systems and Their Applications, Wiley, New York.

3.   Boukerche, A., Rezende, C. & Pazzi, R. (2009). Improving neighbor localization in vehicular ad hoc networks to avoid overhead from periodic messages, pp. 1 –6.

4.   Cheng, L., Bai, F. & Stancil, D. (2009). A new geometrical channel model for vehicle-to-vehicle communications, pp. 1 –4.

5.   Guillemin, E. A., Kalman, R. E., DeClaris, N. & Andersen, J. (1971). Aspects of network and system theory, Holt Rinehart and Winston, New York.

6.   Huang, C.-J., Chuang, Y.-T., Yang, D.-X., Chen, I.-F., Chen, Y.-J. & Hu, K.-W. (2008). A mobility-aware link enhancement mechanism for vehicular ad hoc networks, EURASIP J. Wirel. Commun. Netw. 2008: 1–10.

7.   Kaaniche, H. & Kamoun, F. (2010). Mobility prediction in wireless ad hoc networks using neural networks, Journal of Telecommunications 2(1).

8.   Matolak, D. W., Sen, I. & Xiong, W. (2006). Channel modeling for v2v communications, Mobile and Ubiquitous Systems - Workshops, 2006. 3rd Annual International Conference on, pp. 1 –7.

9.   Maurer, J., Fugen, T., Schafer, T. & Wiesbeck, W. (2004). A new inter-vehicle communications (ivc) channel model, Vol. 1, pp. 9 – 13 Vol. 1.

10.   Molisch, A., Tufvesson, F., Karedal, J. & Mecklenbräuker, C. (2009). A survey on vehicle-to-vehicle propagation channels, Wireless Communications, IEEE 16(6): 12 –22.

11.   Nagel, R. (2010a). Altruistic traffic limits computation in wireless broadcast networks, Proceedings of the Third Internation Conference on Advances in Mesh Networks.

12.   Nagel, R. (2010b). The effect of vehicular distance distributions and mobility on vanet communications, Proceedings of the IEEE Intelligent

Vehicles Symposium.

13. Nagel, R. & Eichler, S. (2008). Efficient and realistic mobility and channel modeling for vanet scenarios using omnet++ and inet-framework, Simutools '08: Proc. of the 1st international conference on Simulation tools and techniques for communications, networks and systems & workshops, ICST, pp. 1–8.

14. Paier, A., Karedal, J., Czink, N., Dumard, C., Zemen, T., Tufvesson, F., Molisch, A. F. & Mecklenbräuker, C. F. (2009). Characterization of vehicle-to-vehicle radio channels from measurements at 5.2 ghz, Wirel. Pers. Commun. 50(1): 19–32.

15. Rezende, C. G., Pazzi, R. W. & Boukerche, A. (2009). An efficient neighborhood prediction protocol to estimate link availability in vanets, MobiWAC '09: Proceedings of the 7th ACM international symposium on Mobility management and wireless access, ACM, New York, NY, USA, pp. 83–90.

16. Rodas, J. & Cascon, C. J. E. (2010). Dynamic path-loss estimation using a particle filter, International Journal of Computer Science Issues 7(3).

17. Samaan, N. & Karmouch, A. (2005). A mobility prediction architecture based on contextual knowledge and spatial conceptual maps, Mobile Computing, IEEE Transactions on 4(6): 537 – 551.

CITATION

CHAPTER 1

P. JANIS, C. YU, K. DOPPLER, C. RIBEIRO, C. WIJTING, K. HUGL, O. TIRKKONEN and V. KOIVUNEN, "Device-to-Device Communication Underlaying Cellular Communications Systems," International Journal of Communications, Network and System Sciences, Vol. 2 No. 3, 2009, pp. 169-178. doi: 10.4236/ ijcns.2009.23019.

CHAPTER 2

R. MACK, M. RIZKALLA, P. SALAMA and M. EL-SHARKAWY, "VLSI Implementation for Low Noise Power Efficiency Cellular Communication Systems," Wireless Sensor Network, Vol. 2 No. 1, 2010, pp. 18-30. doi: 10.4236/wsn.2010.21003.

CHAPTER 3

Anne j. Gilliland-swetland and Greg Kinney, Uses of Electronic Communication to Document an Academic Community: A Research Report, http://journals.sfu.ca/ archivar/index.php/archivaria/article/viewFile/12026/12995.

CHAPTER 4

Ruizhen Han Yong He and Fei Liu, Feasibility Study on a Portable Field Pest Classification System Design Based on DSP and 3G Wireless Communication Technology, doi:10.3390/s120303118.

CHAPTER 5

Md. Shariful Islam, Muhammad Mahbub Alam, Choong Seon Hong and Sungwon Lee, Load-Adaptive Practical Multi-Channel Communications in Wireless Sensor Networks, doi:10.3390/s100908761.

CHAPTER 6

Marco Gramaglia Carlos J. Bernardos and Maria Calderon, Virtual Induction Loops Based on Cooperative Vehicular Communications, doi:10.3390/s130201467.

CHAPTER 7

Youhei Kawamura, Markus Wagner, Hyongdoo Jang, Hajime Nobuhara, Takeshi Shibuya, Itaru Kitahara Ashraf M Dewan and Bert Veenendaa, A Multimedia Data Visualization Based on Ad Hoc Communication Networks and Its Application to Disaster Management, doi:10.3390/ijgi4042004.

CHAPTER 8

Y. Mashhadany, "Design and Implementation of Electronic Control Trainer with PIC Microcontroller," Intelligent Control and Automation, Vol. 3 No. 3, 2012, pp. 222-228. doi: 10.4236/ica.2012.33025.

CHAPTER 9

Huda, K. and Azad, A. (2015) Professional Stress in Journalism: A Study on Electronic Media Journalists of Bangladesh. Advances in Journalism and Communication, 3, 79-88. doi: 10.4236/ajc.2015.34009.

CHAPTER 10

Javier Portela Luis Javier García Villalba Alejandra Guadalupe Silva Trujillo Ana Lucila Sandoval Orozco and Tai-hoon Kim, Extracting Association Patterns in Network Communications, doi:10.3390/s150204052.

CHAPTER 11

P. Visconti, S. D'Amico, A. Baschirotto, et al., "Bidirectional Communication System on Power Line Integrated on Electronic Board for Driving of LED and HID Lamps," Advances in Power Electronics, vol. 2012, Article ID 872383, 10 pages, 2012. doi:10.1155/2012/872383.

CHAPTER 12

Juan Chóliz, Ángela Hernández and Antonio Valdovinos, A Framework for UWB-Based Communication and Location Tracking Systems for Wireless Sensor Networks, doi:10.3390/s110909045.

CHAPTER 13

Robert Nagel and Stefan Morscher (2011). Connectivity Prediction in Mobile Vehicular Environments Backed By Digital Maps, Advanced Trends in Wireless Communications, Dr. Mutamed Khatib (Ed.), ISBN: 978-953-307-183-1, InTech, DOI: 10.5772/15991.

INDEX